13th International Conference on Intelligent Technologies in Logistics and Mechatronics Systems (ITELMS'2020)

Panevezys, Lithuania
1 October 2020

Editors:

Olga Strikulienė
Daiva Žostautienė
László Á. Kóczy
Elvyra Zacharovienė
Ilona Gasiūnienė

ISBN: 978-1-7138-3364-2

Printed from e-media with permission by:

Curran Associates, Inc.
57 Morehouse Lane
Red Hook, NY 12571

Some format issues inherent in the e-media version may also appear in this print version.

Kaunas University of Technology
K. Donelaičio St. 73
44249 Kaunas, Lithuania

Phone: +370 (37) 300 000
Fax: +370 (37) 324 144
Email: ktu@ktu.it

https://itms.ktu.edu

Additional copies of this publication are available from:

Curran Associates, Inc.
57 Morehouse Lane
Red Hook, NY 12571 USA
Phone: 845-758-0400
Fax: 845-758-2633
Email: curran@proceedings.com
Web: www.proceedings.com

13th International Scientific Conference

Intelligent Technologies in Logistics and Mechatronics Systems

ITELMS'2020

October 1st, 2020
Panevėžys, Lithuania

Editors

Olga Strikulienė
Daiva Žostautienė
László Á. Kóczy
Elvyra Zacharovienė
Ilona Gasiūnienė

EDlearning

All papers were reviewed.

www.edlearning.it • congressi@editografica.com

Foreword

The 13th International Scientific Conference Intelligent Technologies in Logistics and Mechatronics Systems – ITELMS'2020 was organized as an integral part of the 3rd Forum on Innovative Technologies and Management for Sustainability that hosted two conferences: 8th International Scientific Conference Changes in Social and Business Environment – CISABE'2020 and 13th International Scientific Conference Intelligent Technologies in Logistics and Mechatronics Systems – ITELMS'2020.

The aim of this 3rd Forum on Innovative Technologies and Management for Sustainability is to provide a worldwide Forum, where the international participants can share their research knowledge and ideas on the latest research findings and map out the directions for future researchers and collaborations.

The main idea of this joint event was to stimulate interdisciplinary approach on sustainability issues since the most advanced, innovative solutions are created by close cooperation between social and technological sciences.

The importance of sustainable development is undeniable, but its implementation into corporate activities is often faced with the problems. Undoubtedly, business can become more sustainable from the perspective of people, planet and profit. However, for many marketers the pursuit of sustainability is considered an uncharted territory with many new and contradictory questions. What are the benefits of incorporating social, ethical and environmental issues into day-to-day decisions making? How can business identify and make the most of the opportunities arising from the new agenda? Striving for sustainability often implies reconsideration of conventional management and marketing principles, stakeholder relationships, the needs for clearer accountability and openness, and higher levels of innovation and co-operation.

The focus of the Forum is on various aspects of contemporary economic, social and technological environments for international business practice(s) and their target market(s). The Forum provided a platform for new forms of cooperation, identification of emerging obstacles and provision of solutions to overcome them.

International Scientific Conference Intelligent Technologies in Logistics and Mechatronics Systems – ITELMS'2020 included the following themes:

- Intelligent Logistics Systems;
- Multi Criteria Decision Making;
- Composites in Infrastructures;
- Intelligent Mechatronics Systems;
- Sensors and Sensing Phenomena;

- Mechanisms of Transport;
- Transport Systems;
- Modern Building Technologies;
- Military Technologies;
- Production Engineering.

This conference emerged as the result of collective efforts and here, we take the opportunity to express our recognition of the efforts and hard work of all the people who have made ITELMS'2020 possible, who have done the hard work in preparation and organization. Through these lines of welcome we would like to especially pay tribute to the Scientific Committee which has assured the quality of the pool of accepted papers, and to the members of the Organizing Committee for their strong motivation and volunteer work, and eventually, to all the people who have directly or indirectly influenced the smooth progress towards the conference. Unfortunately, this is an unusual situation for all scientific community, since COVID-19 has adjusted the ways the Forum is organized. We could not shake hands with each other, but thanks to technology, we could maintain relationship and proceed with discussions. Finally, we would like to express our gratitude to all the authors for contributing their papers to this volume.

Daiva ŽOSTAUTIENĖ

Contents

Forum on Innovative Technologies and Management for Sustainability – ITMS'2020

panevėžys faculty of technologies and business

International Scientific Conference Intelligent Technologies in Logistics and Mechatronics Systems – ITELMS'2020, 1st October, 2020, Panevėžys, Lithuania

Investigation Program for Education Proposed on the Self-Consumption of Power Produced by Small PV Systems

Vytautas Adomavičius[a], Jonas Valickas[b]*

[a]*Kaunas University of Technology, Faculty of Electrical and Electronics Engineering, 48 Studentų St, Kaunas LT-51367, Lithuania*
[b]*Kaunas University of Technology, Panevėžys Faculty of Technologies and Business, 33 Nemuno St, Panevėžys LT-37164, Lithuania*

Abstract

The aim of this paper is to present a short review on the self-consumption of electric power produced in small-scale PV power plants, disseminate some basic information about this technology, and to offer for students an experimental laboratory work program that would enable them to acquire knowledge and skills in the field of self-consumption of electricity produced by solar power plants with micro inverters. Advantages and shortages of self-consumption are described. Dependence of power flows between the output of micro inverters, electric load and electric grid of power system are investigated and explained depending on the solar irradiance variations, time and capacity of electric load. The results of the research are presented in tables and in the form of curves. Results of the investigation are summarised in conclusions and discussed.

Keywords: solar energy; small-scale photovoltaic power plants; grid-connected PV system; micro inverters; self-consumption of power; anti-islanding protection; education in the field of PV power plants.

Introduction

Self-consumption of energy from the grid connected small-scale photovoltaic system is a relatively new way of using solar energy when in some cases it is not necessary to use meters for the electricity produced. This method is increasingly used for the integration of low-power solar power plants with micro inverters into the electricity grid in various countries. Potential users of this technology quite often do not need to have any project for installation of this system in their house, apartment or company, do not need to coordinate any contract and sign it with the electric power system authorities, because these small power plants use safe smart micro inverters and all equipment works reliably and has multi-year warranties. The electricity produced by a small power plant is primarily consumed by the house (apartment, private institution), but if it is not used, the power flows to the external electricity network. When the

* Corresponding author. Tel.: + 370 656 24494.
E-mail address: jonas.valickas@ktu.lt

DX20F0001

mains voltage is disconnected (or disappears automatically due to other reasons), the micro inverters momentarily cut off the power supply to the mains. Such protection of the mains against unexpected voltage surges, which is dangerous for service personnel, is called anti-islanding protection.

The first microinverters were developed by the US company Ascension Technology (1991), and the first commercially successful microinverters were introduced in large quantities from 2008 onwards. Since then, microinverters have been successfully sold worldwide. In the beginning, microinverters were relatively much more expensive than inverters for large solar power plants. The price index of microinverters was about 2–3 times higher. But over time, their prices have dropped quite rapidly, as have dropped prices of solar modules and all solar power systems. Currently, the price index of microinverters is about 45 ct/W, and of high power solar inverters (string inverters – branch inverters) – about 30–35 ct/W. Such a small difference in the price indices of inverters is not significant, as the capacity of solar power plants with microinverters is low and no major financial problems arise as a result. In addition, over time, the spread between their price indices narrows quite rapidly, so it can be expected that these indices will probably level off in a few years.

Solar power plants, in which each photovoltaic module has its own low-power microinverter, are characterized by higher energy efficiency, greater reliability, and greater economic efficiency. Compared to traditional grid-connected solar power plants, which have one powerful inverter (or several powerful inverters when the power of the solar power plant is very high), solar power plants with microinverters are easier to design and much easier to install, maintain or dismantle. Powerful inverters in traditional high-power solar power plants with mutual maximum power point trackers (MPPTs) have a large number of solar modules that are often in unequal conditions (for example, some modules are in the shade). In such cases, it was quite often possible to achieve the maximum possible capacity of the solar module only in part, because then, depending on the working conditions, the efficiency of the MPPT decreases. More efficient operation of PV systems with microinverters partially offsets the higher price index of microinverters. The main advantages of the solar power plants with microinverters compared to the solar power plants with common inverters of high capacity for all or a large part of the modules are the following:

- the efficiency of the whole solar power plant and the amount of electricity produced increase because the MPPT works for a single module or a very small group of modules;
- the reliability of this type of solar power plant of higher power increases, because when one microinverter fails, only a small part of the whole solar power plant does not work, the rest of it works;
- simplification of the installation, maintenance, repair and dismantling of the solar power plant at the end of the power plant's service life;
- the installation, maintenance and dismantling at the end of life of the solar power plant becomes simpler;
- reduces power and energy losses in a solar power plant and increases the economic efficiency of its operation, shortens its payback time.

Self-consumption policies regarding the PV power plants are different in various countries [1–5]. Also different schemes are used for the metering of self-consumed electricity. In some countries small-scale PV power plants can be installed without a power meter. Information sources on the topic of self-consumption of electricity can be found in a number of useful scientific publications, which examine the possibilities of improving this technology [6, 7], its operation under various load scenarios [8], its use for charging of EV batteries [9], the use of electric power storage for improving of this system operation [10, 11], the economic evaluation of this technology [12, 13], solves the problems of its performance improvement [14] and optimization [15].

Nomenclature

E	the solar irradiance, W/m^2;
P_{rl}	the rated load of electric bulbs, W;
P_{al}	the actual load of electric bulbs at the actual operating voltage, W;
P_m	the power produced by the micro inverters at current solar irradiance, W;
P_g	the power supplied to the public grid or taken from the grid, W;
MI	micro inverter;
MPPT	maximum power point tracker;

PV	photovoltaic power plant;
SM	solar module.

1. Object of research and methods

The object of our research is a small-scale grid connected photovoltaic system with microinverters. The aim of the research was to investigate the operating modes and power flux distributions in such a system depending on the changes of the solar irradiance and the electrical load of this PV power plant. Currently, one to four solar modules are usually connected to one micro inverter in such PV systems, as shown in Figure 1 and 2.

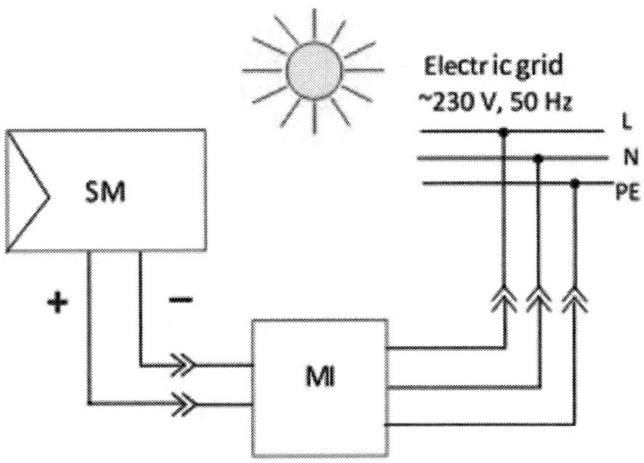

Fig. 1 Connecting a micro inverter for one SM to the power grid

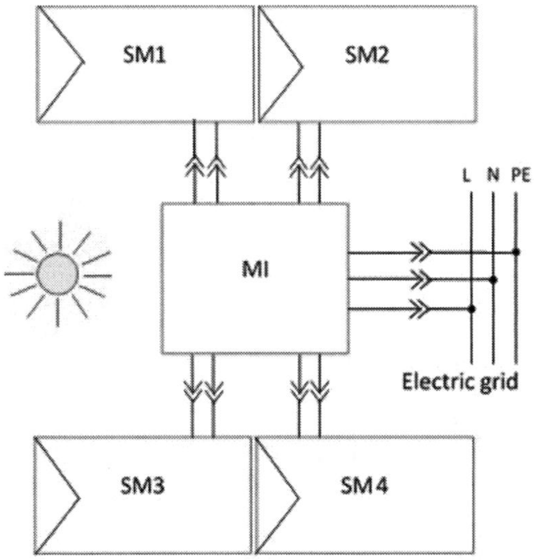

Fig. 2 Connecting a micro inverter for four SM to the power grid

Experimental PV power plant and its operating modes. The pilot PV power plant has two polycrystalline silicon solar modules mounted on the roof of the building and oriented to the south with a slope of 38 degrees to the horizontal plane. Rated power of solar modules – 218 Wp. This PV power plant also has two micro innverters (MI), the load unit, and measuring devices used to measure fluxes of power in the circuits of the PV system output. Scheme of electric connections of small PV system with micro inverters is presented in Figure 3. The most important parameters of MI used in the experimental stand are presented in Table 1. The main part of the experimental PV system was mounted in laboratory on the stand intended for the investigations of self-consumption is shown in Figure 4.

Table 1. The main data of micro inverters

MI input data (DC)		MI output data (AC)			
Parameter	Value	Parameter	Value	Parameter	Value
Recommended input power range	180–360 W	Maximum power	300 W	Power factor	> 0.99
Maximum input DC voltage	54 V	Maximum current	1.36 A	Total harmonic distortion	< 3 %
MPPT voltage range	24–42 V	Nominal grid voltage	220/230 V	Nighttime power consumption	< 100 mW
Full load MPPT voltage range	24 V	Nominal grid frequency	50/60 Hz	Peak micro inverter efficiency	≤ 95.6 %
MPPT disconnection voltage	18 V	Grid voltage range	165–256 V	CEC efficiency	95 %

MI input data (DC)		MI output data (AC)			
Parameter	Value	Parameter	Value	Parameter	Value
Maximum DC short circuit current	9.5 A	Grid frequency range	47–51.5 Hz	MPPT efficiency	99.9 %
Maximum input current	15 A	Dimensions	163 × 163 × 27	Max number of units in branch	20

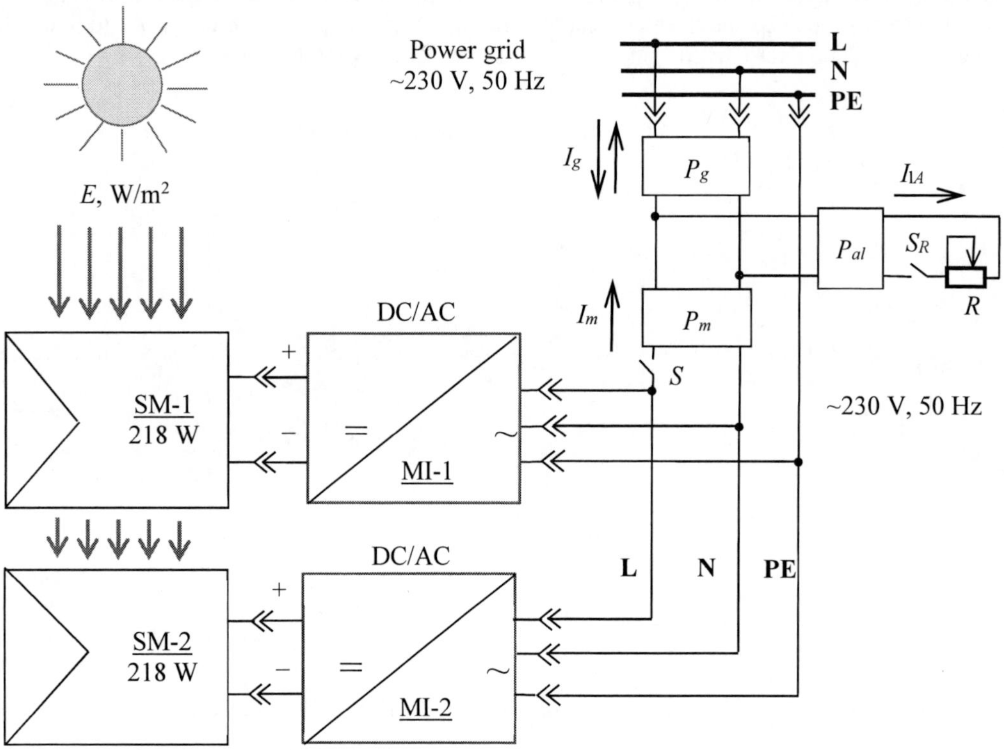

Fig. 3. Scheme of electric connections of small PV system with micro inverters

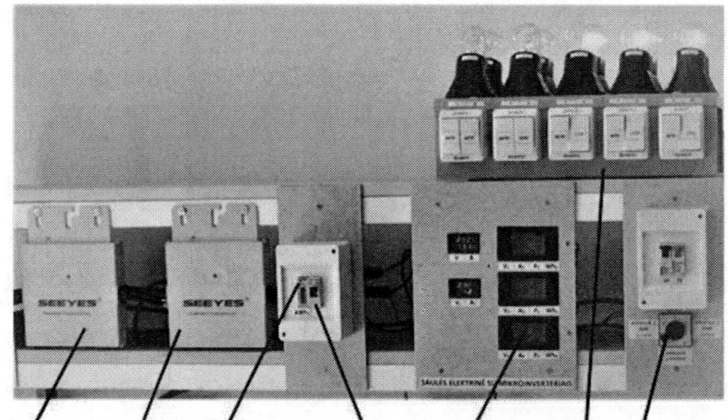

Fig. 4. The main part of the PV system mounted on the stand for investigations: 1, 2 – micro inverters; 3 – electric energy meter; 4 – switch S; 5 – electric power meters P_m, P_g, P_{al}; 6 – load unit R; 7 – switch S_R

The load unit 6 is equipped with light bulbs for illumination with a rated wattage of 230 V. As the output voltage of the micro inverters is higher than 230 V (about 237 V), the actual wattage of the bulbs is higher than rated. Using switch *S*, the anti-islanding protection of the micro inverter can be checked. When the switch 4 is turned off, the micro inverters must stop generating the output voltage immediately. After this signal lights of the micro inverters 1 and 2 start flashing red (flashing light is green when the inverters are supplying power to the mains).

2. Results of the investigation

This section presents the results of an experimental investigation of a small-scale PV power plant that is suitable for operation in self-consumption mode because its micro inverters have anti-islanding protection. These results could obviously be shown by the visualized electric power flows in the form of curves between the output of micro inverters, electric load and electric grid of power system. Therefore we researched the power flows from micro inverters, to (or from) the electric grid of power system and to the load depending on the three variables: solar irradiance, time, and capacity of electric load. Three experiments have been performed and the following dependencies have been identified:

1. Power flows in the investigated PV system with micro inverters on the time at cloudy sky with often clarifications when electric load is disconnected and solar irradiance is varying in range 600–250 W/m^2.
2. Power flows in the investigated PV system with micro inverters with varying electric load when solar irradiance is quasistable (varying in range 770 W/m^2 ± 10 W/m^2).
3. Power flows in the investigated PV system with micro inverters on the time at constant load and cloudy sky when solar irradiance is varying in range 830–585 W/m^2.

Results of the mentioned above investigations are presented below in Table 2, 3 and 4. A more obvious visualized version of the obtained results of this study is presented in Figure 5, 6 and 7.

Table 2. Dependence of power flows in the investigated PV system with micro inverters on the time at cloudy sky with often clarifications

t, min	P_{rl}, W	P_{al}, W	P_m, W	P_g, W	Balance of powers $P_m - P_{al} \pm P_g = 0$	Explanations
	14 September, 2020					
	At the cloudy sky with often clarifications and variations of irradiance $E \approx$ 600–250 W/m^2					
1	Electric load is discon-nected	Electric load is discon-nected	196.7	-195.1	1.6	Power output P_m from the micro inverters MI-1 and MI-2 and the power P_g supplied into the power grid differ very little because of the power losses in the connecting wires and contacts in the scheme used for investigation. The losses make up about 1 % of power generated by the micro inverters. Therefore, the curves P_m and P_g practically coincide (Fig. 5).
2			233.3	-231.6	1.7	
3			224.6	-223.0	1.6	
4			200.3	-198.7	1.6	
5			155.8	-154.2	1.6	
6			133.4	-132.0	1.4	
7			154.6	-153.1	1.5	
8			110.1	-108.8	1.3	
9			113.8	-112.5	1.3	
10			146.5	-145.0	1.5	
11			171.2	-169.6	1.6	
12			130.4	-129.0	1.4	
13			115.8	-114.5	1.3	
14			135.7	-134.3	1.4	
15			160.2	-158.7	1.5	
16			195.8	-194.2	1.6	
17			235.4	-233.8	1.6	
18			241.5	-239.8	1.7	
19			250.9	-249.2	1.7	

t, min	P_{rl}, W	P_{al}, W	P_m, W	P_g, W	Balance of powers $P_m - P_{al} \pm P_g = 0$	Explanations
	14 September, 2020					
	At the cloudy sky with often clarifications and variations of irradiance $E \approx 600–250$ W/m^2					
20			150.5	-149.0	1.5	
21			127.7	-126.2	1.5	

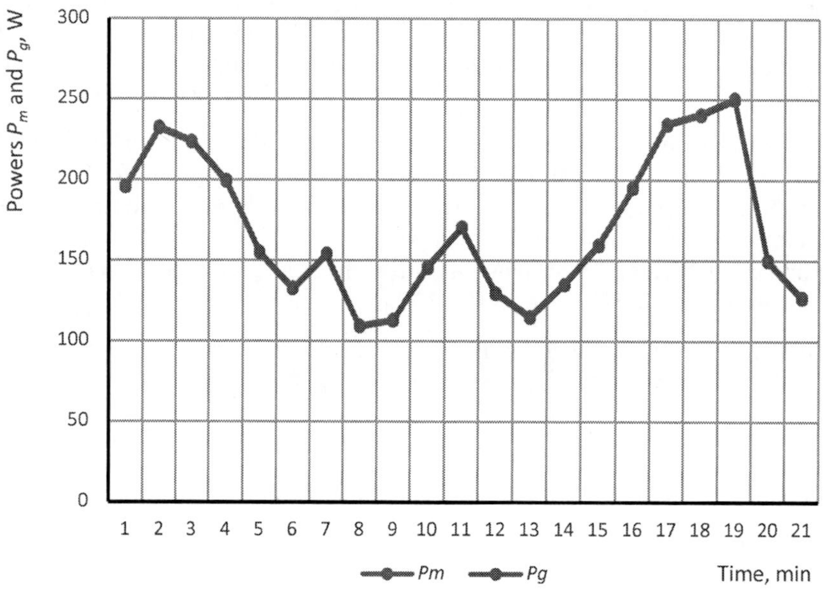

Fig. 5. Dependence of power P_m generated by the micro inverters and power P_g supplied into the grid on the time at cloudy sky when variations of irradiance are in range E ≈ 600–250 W/m^2

Table 3. Power distribution in the investigated PV system with micro inverters with varying electric load when solar irradiance is quasistable

No.	P_{rl}, W	P_{al}, W	P_m, W	P_g, W	Balance of powers $P_m - P_{al} \pm P_g = 0$	Explanations
	15 September, 2020					
	At the clear sky and quasi stable solar irradiance $E \approx 770$ W/m^2 (±10 W/m^2)					
1	0	0	249.0	-247.3	1.7	Power from micro inverters is more than required from the load, therefore a part of power generated by the MI flows into the mains.
2	60	65.8	249.2	-181.9	1.5	
3	120	128.3	249.0	-119.4	1.3	
4	180	190.8	248.3	-56.4	1.1	
5	240	253.2	248.2	5.8	-0.8	All the power generated by the micro converters is used to meet the load needs, but when they start to be insufficient, more and more energy is taken from the grid to sport the small PV power plant.
6	300	315.5	248.7	67.4	-0.6	
7	340	379.8	248.7	31.6	-0.6	
8	380	442.7	248.5	194.8	-0.5	
9	405	465.9	247.6	218.7	-0.4	
10	430	491.5	247.9	244.0	-0.4	

No.	P_{rl}, W	P_{al}, W	P_m, W	P_g, W	Balance of powers $P_m - P_{al} \pm P_g = 0$	Explanations
			15 September, 2020			
	At the clear sky and quasi stable solar irradiance $E \approx 770$ W/m^2 (±10 W/m^2)					
11	455	517.6	247.8	270.2	-0.4	

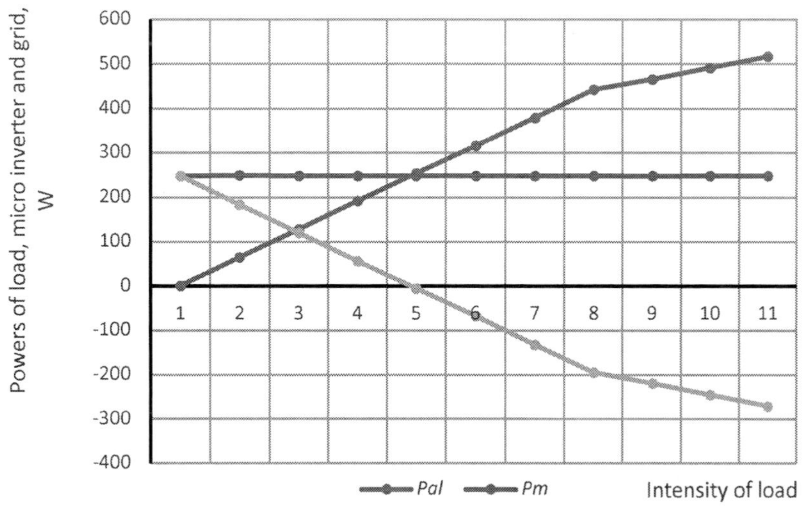

Fig. 6. Distribution of power flows in the investigated PV system with micro inverters dependently on the increasing intensity of electric load when solar irradiance is quasistable

Table 4. Dependence of power flows in the investigated PV system with micro inverters on time at constant load and cloudy sky

t, min	P_{rl}, W	P_{al}, W	P_m, W	P_g, W	Balance of powers $P_m - P_{al} \pm P_g = 0$	Explanations
			16 September, 2020			
	At the cloudy sky with often clarifications and variations of irradiance $E \approx 585$–830 W/m^2					
1	255	266.7	343.4	-75.0	1.7	Solar irradiance during the investigation was sufficient for the set rated load of 255 W throughout the all experimental period, and the cumulative power generated by both micro inverters (P_m – the red curve) was higher than the required load (P_{al} – the blue curve). The power P_m is not constant because the solar irradiance was not constant during the experiment. Unused power of the micro inverters was supplied to the mains (P_g – the green curve). The power from the grid was not used for the load feeding in this case.
2		266.8	351.2	-82.7	1.7	
3		265.5	339.7	-72.8	1.3	
4		265.6	338.9	-72.0	1.3	
5		265.2	329.4	-62.8	1.4	
6		267.0	352.8	-84.4	1.4	
7		267.2	358.8	-90.3	1.3	
8		267.2	364.5	-96.0	1.3	
9		267.3	362.4	-93.8	1.3	
10		266.7	349.4	-81.3	1.4	
11		265.1	329.5	-62.9	1.5	
12		265.3	325.5	-58.7	1.5	
13		265.8	321.3	-54.2	1.3	
14		267.6	369.1	-100.1	1.4	
15		265.4	313.4	-46.8	1.2	
16		265.2	308.0	-41.5	1.3	
17		265.8	305.9	-38.7	1.4	
18		265.5	303.8	-36.9	1.4	

t, min	P_{rl}, W	P_{al}, W	P_m, W	P_g, W	Balance of powers $P_m - P_{al} \pm P_g = 0$	Explanations
			16 September, 2020			
At the cloudy sky with often clarifications and variations of irradiance $E \approx 585–830$ W/m^2						
19		263.7	284.2	-19.2	1.3	
20		265.0	316.7	-50.2	1.5	
21		266.9	349.7	-81.2	1.6	

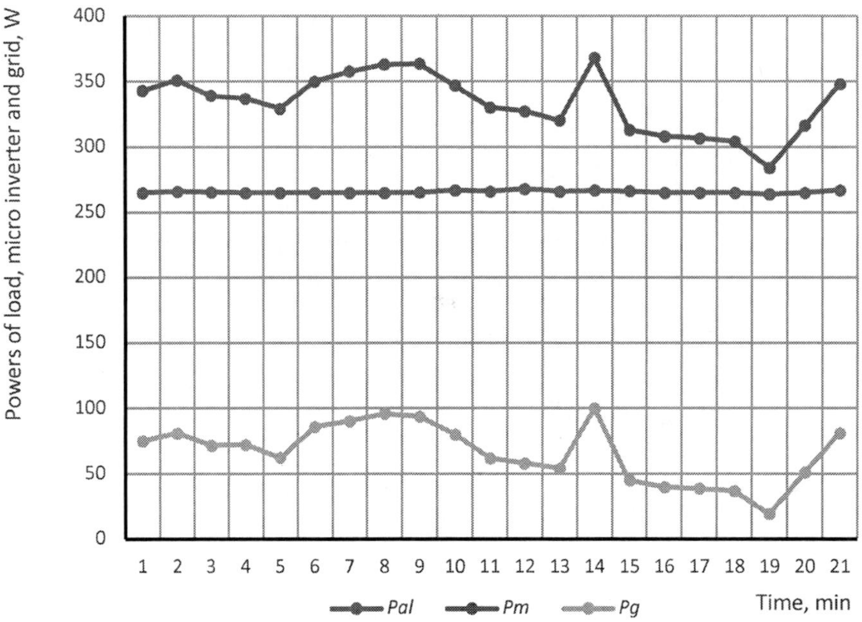

Fig. 7. Distribution of power flows in the investigated PV system with micro inverters at constant rated power of load (255 W) dependently on the time

Conclusions

1. The technology of self-consumption of power produced in RES-based power plants (mostly power of small-scale PV and wind power plants are used for self-consumption) further simplifies the installation and use of renewable power plants – such power plants are becoming available to almost all citizens in the same way as other household electrical appliances.
2. Self-consumption of electric power produced in PV power plants are highly recommended for use in cases when the building uses a lot of electricity during the day when the sun is shining.
3. Self-consumption can also be used in residential buildings, including apartment buildings, if they have sufficient electricity demand during the day, such as water heating, space heating or air conditioning, or other significant electricity consumers.
4. Rational sizing and exploitation of RES-based power plants with embedded technology of power self-consumption can significantly reduce the bills for electricity in buildings.
5. The information presented in this article may be useful for students of high schools, colleges or universities who want to acquire knowledge and skills in the field of self-consumption of electricity produced by solar power plants with micro inverters or who want to perform a similar experimental research.

References

[1] Masson, G., Briano, J. I., Baez, M. J. Review and analysis of PV self-consumption policies [Internet]. *Report IEA-PVPS T1-28:2016.* 82 p. Available from: https://iea-pvps.org/wp-content/uploads/2020/01/IEA-PVPS_-_Self-Consumption_Policies_-_2016_-_2.pdf (accessed on 11-08-2020).

[2] Jäger-Waldau, A., Bucher, Ch., Frederiksen, K. H. B., et al. Self-consumption of electricity produced from PV systems in apartment buildings – comparison of the situation in Australia, Austria, Denmark, Germany, Greece, Italy, Spain, Switzerland and the USA [Internet]. *The 7th World Conference on Photovoltaic Power Conversion (WCPEC-7)*, 10–15 June, 2018. Hilton, Waikaloa, HI, USA. Available from: https://www.researchgate.net/publication/325988881 (accessed on 11-08-2020).

[3] Jäger-Waldau, A., Adinolfi, G., Batlle, A., et al. Self-consumption of electricity produced with photovoltaic systems in apartment buildings [Internet]. *Proceedings of the IEEE PVSC-47* (virtual meeting). 15 June – 21 August, 2020; 16 p. Available from: https://www.researchgate.net/publication/342261164 (accessed on 11-08-2020).

[4] Prol, J. L., Steininger, K. W. Photovoltaic self-consumption regulation in Spain: profitability analysis and alternative regulation schemes [Internet]. *Energy Policy*, 2017; 108:742–754. Available from: http://dx.doi.org/10.1016/j.enpol.2017.06.019 (accessed on 20-08-2020).

[5] Villara, C. H., Neves, D., Silva, C. A. Solar PV self-consumption: an analysis of influencing indicators in the Portuguese context [Internet]. *Energy Strategy Reviews*, 2017; 18:224–234. Available from: https://doi.org/10.1016/j.esr.2017.10.001 (accessed on 20-09-2020).

[6] Luthander, R. *Improved Self-Consumption of Photovoltaic Electricity in Buildings. Thesis* [Internet]. 2016; 83 p. Available from: https://www.researchgate.net/publication/302587113 (accessed on 11-09-2020).

[7] Luthander, R., Widén, J., Nilsson, D., Palm, J. Photovoltaic self-consumption in buildings: a review [Internet]. *Applied Energy*, 2015; 142:80–94. Available from: http://dx.doi.org/10.1016/j.apenergy.2014.12.028 (accessed on 20-09-2020).

[8] Maranda, W. Analysis of self-consumption of energy from grid-connected photovoltaic system for various load scenarios with short term buffering [Internet]. *SN Applied Sciences*, 2019; 1(406):10. Available from: https://doi.org/10.1007/s42452-019-0432-5 (accessed on 22-08-2020).

[9] Fachrizal, R., Munkhammar, J. Improved photovoltaic self-consumption in residential buildings with distributed and centralized smart charging of electric vehicles [Internet]. *Energies*, 2020; 13:1153. Available from: doi:10.3390/en13051153 (accessed on 22-08-2020).

[10] Vieira, F. M., Moura, P. S., de Almeida A. T. Energy storage system for self-consumption of photovoltaic energy in residential zero energy buildings [Internet]. *Renewable Energy*, 2017; 103:308–320. Available from: http://dx.doi.org/10.1016/j.renene.2016.11.048 (accessed on 11-08-2020).

[11] Castillo-Cagigal, M., Caamano-Martin, E., Matallanas, E., et al. PV self-consumption optimization with storage and active DSM for the residential sector [Internet]. *Solar Energy*, 2011; 85:2338–2348. Available from: doi:10.1016/j.solener.2011.06.028 (accessed on 11-08-2020).

[12] Camilo, F. M., Castro, R., Almeida, M. E., Pires, V. F. Economic assessment of residential PV systems with self-consumption and storage in Portugal [Internet]. *Solar Energy*, 2017; 150:353–362. Available from: http://dx.doi.org/10.1016/j.solener.2017.04.062 (accessed on 20-08-2020).

[13] Lang, T., Ammann, D., Girod, B. Profitability in absence of subsidies: a techno-economic analysis of rooftop photovoltaic self-consumption in residential and commercial buildings [Internet]. *Renewable Energy*, 2016; 87:77–87. Available from: http://dx.doi.org/10.1016/j.renene.2015.09.059 (accessed on 22-08-2020).

[14] Castillo-Cagigal, M., Gutiérrez, A., Monasterio-Huelin, F., et al. A semi-distributed electric demand-side management system with PV generation for self-consumption enhancement [Internet]. *Energy Conversion and Management*, 2011; 52:2659–2666. Available from: doi:10.1016/j.enconman.2011.01.017 (accessed on 11-08-2020).

[15] Martín-Chivelet, N., Montero-Gómez, D. Optimizing photovoltaic self-consumption in office buildings [Internet]. *Energy and Buildings*, 2017; 150:71–80. Available from: http://dx.doi.org/10.1016/j.enbuild.2017.05.073 (accessed on 11-08-2020).

Forum on Innovative Technologies and Management for Sustainability – ITMS'2020

ktu 1922 panevėžys faculty of technologies and business

International Scientific Conference Intelligent Technologies in Logistics and Mechatronics Systems – ITELMS'2020, 1st October, 2020, Panevėžys, Lithuania

Research of Possibilities for Reduction of Heat Losses in Apartment Buildings

Vytautas Adomavičius[a], Jonas Valickas[b]*

[a]*Kaunas University of Technology, Faculty of Electrical and Electronics Engineering, 48 Studentų St, Kaunas LT-51367, Lithuania*
[b]*Kaunas University of Technology, Panevėžys Faculty of Technologies and Business, 33 Nemuno St, Panevėžys LT-37164, Lithuania*

Abstract

Goal of this paper is to present results of the research regarding the ways of enhancing energy efficiency in a typical renovated apartment building. The main emphasis in this research is put on the efficiency of domestic hot water preparation but possibilities to reduce heat energy consumption for the space heating are analysed and discussed as well. Thermal energy wastage in the process of DHW production was checked by the performing thermal energy consumption monitoring in the DHW pipeline in the section from the heating unit to the inlets to the apartments. The measurements also were performed by means of the heat energy meters installed in the space heating pipelines. Monitoring of the building's thermal energy system showed that after the renovation, the space heating system works very efficiently. Losses of heat energy are not high in this system. However, the DHW system has a very high heat loss, consuming up to three times more heat than would be required. This is also reflected in the bills received by apartment owners for space heating and DHW – space heating is very cheap and DHW is very expensive. The monitoring of the thermal energy system of the building allowed determining the extent of heat losses in the DHW system. Simple and easy-to-implement measures have been proposed to reduce these losses.

Keywords: energetically efficient buildings; heat energy; possibilities to enhance energy efficiency of domestic hot water supply and space heating.

Introduction

Energy saving in all possible ways and the fastest possible transition to the use of renewable energy sources in all sectors of the economy has become very important in the 21st century due to the rapidly accelerating global climate crisis and very intensive pollution of environment. Energy systems in buildings are one of the areas where much can be done in order to increase energy efficiency, reduce emissions of the greenhouse gases (GHG) and other pollutants. People rightly expect that reduced pollution in the energy, industrial, transport and other sectors of the economy will

* Corresponding author. Tel.: + 370 656 24494.
E-mail address: jonas.valickas@ktu.lt

DX20F0002

improve their health, increase life expectancy and reduce healthcare costs, as the concentration of pollutants in the air they are now forced to breathe will decrease. Smart people do not start smoking cigarettes or can quit it very easy, but it is impossible to quit "smoking" industrial and transport pollutants, which exist almost everywhere. Currently rather often pollutants are released into the Earth's atmosphere without any pollution taxes, which only the most advanced and responsible countries have introduced before decades of years. The highest concentrations of pollutants are in cities and around the highways. Where possible, such places should be avoided. Detailed information on the intensity of air pollution in the United States and its harmful effects on lung health alone is provided in an annual report 2020 issued by the Lung Association of this country [1].

The negative effects of pollutants from fossil fuel combustion on human health, species of fauna and flora are not the only damage to our planet. Many difficult problems and huge material losses are caused by climate change, which is now entering a phase of climate crisis. Key indicators of climate change correlate with increases in GHG concentration in the Earth's atmosphere. Key facts about this relationship can be found in the article "World Scientist's Warning of a Climate Emergency" [2] of the journal BioScience, published by Oxford University. Joining the article's warning about the dangers of climate change, the article has been signed by more than 20 000 signatory scientists from 153 countries.

Apartment buildings and buildings in general are only one of many possible branches of economy where substantial progress in energy efficiency and the transition from fossil fuel energy to renewable energy can be done now without significant delay. Our review of the scientific literature on efficient and sustainable energy usage in buildings has shown that the potential of renewable energy in buildings has been explored for more than 20 years. Various energy systems and measures to increase the efficiency of RES-based energy systems suitable for exploitation in buildings are analyzed in the publications [3–5]. The publications [6–9] are devoted to the problems of optimization of renewable energy systems in buildings. The publication [10] addresses a very relevant problem – the balance between saving energy in a building and capacity of heat supplying system for the same building. Energy saving in buildings paves the way to the successful implementation of the low cost RES-based energy systems for the needs of the buildings.

Authors of this paper are also interested in solving of the energy supply problems related with buildings and suggest to rely first of all on the efficient renewable energy sources that are free, environmentally friendly and almost universally available. In many cases it could be hybrid solar and wind energy systems together with power and heat storage facilities. The grid of the power system can also be used as a backup energy source. Solar and wind energy can be used in one family houses [11], in apartment buildings [12] and in any other type of buildings.

This article investigates the possibilities of reducing in a typical apartment building in Kaunas. The space heating system installed in Soviet times in this building was very successfully renovated a few years ago. Costs for heating of apartments decreased by about 2–3 times after the renovation. However, the domestic hot water system is operating very inefficiently and consumes a lot of thermal energy. Thermal energy meters were installed in the heating unit and also at the hot water inlets to the apartments on purpose to find out the causes and amounts of thermal energy wastage.

Nomenclature

C_{cw}	the price of 1 m^3 of cold running water, €/m^3;
C_{he}	the price of the heat energy used for water heating, €/kWh;
C_{hw}	the price of 1 m^3 of running water heated up to 55 °C (including the price of cold water), €/m^3;
C_{wh}	the average price of the water heating from the temperature of running water (6–9 °C) up to 55 °C, €/kWh;
C_{wkh}	the price of the hot water in the supply system of the building keeping hot (55 °C) for one apartment, €;
E_{wh}	the average heat energy consumed per month for heating of 1 m^3 running water up to 55 °C, kWh/m^3;
E_{wkh}	the average heat energy consumed per month for 1 m^3 of DHW keeping hot in the pipe system, kWh/m^3;
E_a	the actual heat energy consumed per month for 1 m^3 of DHW (for heating and keeping hot), kWh/m^3;
V_{hw}	the consumed volume of the hot water by the apartment per month, m^3;
DHW	domestic hot water;

HWP	hot water circulation pump;
GHG	greenhouse gases;
RES	renewable energy sources.

1. Objects of research and methods

The main object of our research are the thermal energy losses in the domestic hot water system of the apartment building in campus of Kaunas University of Technology built in the year 1987. Technical characteristic of the building is given in Table 1.

Table 1. Technical parameters of the researched apartment building

No.	Parameters of apartment building	Units	Values	Notes
1.	Number of apartments	–	75	
2.	Number of stairwells	–	7	
3.	Number of stories	–	6	
4.	Number of dwellers	–	230–250	Has trend to decrease
5.	Total heated area	m^2	6 750	
6.	Total heated space	m^3	16 500	
7.	Average annual consumption of DHW	m^3	3 600	
8.	Average monthly consumption of DHW	m^3	300	
9.	Average daily consumption of DHW	m^3	10	
10.	Price of 1 m^3 of cold running water from the public water supply	€/m^3	1.56	During the last 10 years
11.	Price of heat energy used for space heating and DHW	ct/kWh	3.69–6.04	Depends on the fuel
12.	Price of the DHW in the investigated apartment building	€/m^3	4.80–3.16	Depends on the month
13.	Rated temperature of the DHW in the apartment building	°C	55 ± 2	
14	Average amount of heat energy for production of 1 m^3 DHW	kWh/m^3	45.95	During the last 4 years
15.	Annual consumption of heat energy for DHW in 2017	MWh	508.75	
16.	Average annual consumption of heat energy for space heating	kWh/m^2	40	Depends on weather meteorological parameters

Causes and amounts of thermal energy wastage in the process of DHW production were determined by performing constantly operating thermal energy consumption monitoring in the DHW pipeline in the section from the heating unit up to the inlets to the apartments. The measurements were performed by means of the heat energy meters installed in the pipelines of DHW and space heating. Official bills for the consumed DHW and for space heating are prepared periodically for the owners of the apartments according to the readings of these heat energy meters, which are read every month.

All calculations and investigations in this paper are based on the data presented in the official bills submitted to the owners of the building's apartments. They include the following data:

- number of the month of the year;
- price of the heat energy for that month, €/kWh;
- price of hot water per cubic meter for that month, €/m^3;
- volume of hot water used in the apartment per that month, m^3;
- price of the total volume of hot water consumed in the apartment per that month, €;
- price of the thermal energy consumed for the total volume of hot water consumed in the same apartment per that month in order to keep the water hot in the hot water pipeline of the building (the hot water pipeline of the building comprises all pipes from the inlet of the district heating water up to the inlets of hot water into apartments).

We are interested in the amounts of thermal energy used in the hot water production process, which are not presented in the official bills submitted to the owners of the building's apartments. However, they can be calculated on the basis of data provided in the official bills given for every apartment of the building. The actual and believable amounts of heat energy losses can only become clear when we find out how much heat energy is used to produce hot water and how much energy is needed to keep the water hot (55 °C) and circulating in the building's domestic hot water pipeline.

The thermal energy consumed for the production of hot water from the cold running water for the certain apartment mentioned in the bill can be calculated in this way:

$$E_{wh} = (C_{hw} - C_{cw}) \ / \ C_{he}. \qquad (1)$$

The heat energy required to maintain the rated temperature of hot water in the hot water pipeline (55 °C) of the building from the heating unit to the hot water inlet of the certain apartment mentioned in the bill can be calculated in this way:

$$E_{wkh} = (C_{wkh} \ / \ C_{he}) \ / \ V_{hw}. \qquad (2)$$

The actual energy consumption E_a associated with the preparation of the DHW and its maintenance at the specified temperature is obtained by summing the energy consumptions calculated according to the given above formulas (1) and (2):

$$E_a = E_{wh} + C_{wkh}. \qquad (3)$$

The most important parameters of DHW produced in the investigated apartment building in the years 2016–2018 are presented in Tables 2–4. Prices in the tables related to hot water production are taken from the official bills given to the apartments of the building. The amounts of DHW-related energy in these tables were calculated using the formulas (1), (2) and (3).

Table 2. The main parameters of the DHW for the year 2016

2016	C_{cw}	C_{he}	C_{hw}	V_{hw}	C_{wh}	C_{wkh}	E_{wh}	E_{wkh}	E_a
Month	€/m³	€/kWh	€/m³	m³	€	€	kWh/m³	kWh/m³	kWh/m³
1	1.56	0.059	4.25	3.47	14.76	9.83	45.59	48.01	93.60
2		0.0586	3.98	1.98	7.89	9.77	41.30	84.20	125.50
3		0.0604	4.80	1.85	8.89	10.06	53.64	90.03	143.67
4		0.0525	3.94	2.27	8.96	8.64	45.33	72.50	117.83
5		0.0460	3.78	1.88	7.12	6.86	48.26	79.33	127.59
6		0.0428	3.45	2.11	7.29	5.39	44.12	59.63	103.75
7		0.0435	3.48	3.04	10.59	5.03	44.15	38.04	82.19
8		0.0402	3.30	2.46	8.13	5.51	43.26	55.69	98.95
9		0.0402	3.31	2.28	7.56	5.58	43.51	60.85	104.36
10		0.0437	3.48	2.29	7.98	7.19	43.94	71.85	115.78
11		0.0514	3.55	2.23	7.93	8.46	38.72	73.81	112.52
12		0.0534	4.00	2.51	10.05	8.76	45.69	65.36	111.05
					Averages		**44.79**	**66.61**	**111.40**

Comparison of average heat energy parameters related with DHW production in the years from 2016 up to 2018 are given in Table 5. The data in this table show that the actual amounts of thermal energy E_a used to produce DHW and maintain its rated temperature are approximately 2.5 to 3 times higher than thermal energy E_{wh} necessary to heat cold running water to the rated temperature 55 °C. The average value of heat energy overrun is 2.73 times.

Table 3. The main parameters of the DHW temperature for the year 2017

2017	C_{cw}	C_{he}	C_{hw}	V_{hw}	C_{wh}	C_{wkh}	E_{wh}	E_{wkh}	E_a
Month	€/m^3	€/kWh	€/m^3	m^3	€	€	kWh/m^3	kWh/m^3	kWh/m^3
1	1.56	0.0557	4.10	2.00	8.21	9.16	45.60	82.23	127.83
2		0.0557	4.42	1.63	7.21	9.16	51.35	100.89	152.24
3		0.0523	3.81	1.34	5.12	8.60	43.02	122.71	165.73
4		0.0464	3.72	1.71	6.37	7.63	46.55	96.16	142.72
5		0.0449	3.86	1.69	6.53	8.09	51.22	106.60	157.81
6		0.0471	3.89	1.76	6.86	7.47	49.50	90.17	139.67
7		0.0463	3.89	2.67	10.4	6.75	50.28	54.55	104.83
8		0.0467	3.86	1.77	6.84	7.46	49.24	90.24	139.48
9		0.0469	3.84	1.68	6.46	7.82	48.56	99.15	147.71
10		0.0448	3.65	1.53	5.60	7.25	46.65	105,77	152.42
11		0.0519	3.93	1.71	6.73	8.13	45.66	91.61	137.27
12		0.0552	3.78	1.84	6.97	8.93	40.22	87.92	128.14
					Averages		**47.32**	**94.00**	**141.32**

Table 4. The main parameters of the DHW temperature for the year 2018

2018	C_{cw}	C_{he}	C_{hw}	V_{hw}	C_{wh}	C_{wkh}	E_{wh}	E_{wkh}	E_a
Month	€/m^3	€/kWh	€/m^3	m^3	€	€	kWh/m^3	kWh/m^3	kWh/m^3
1	1.56	0.0603	4.50	2.62	11.8	9.76	48.76	61.78	110.53
2		0.0597	4.35	1.65	7.19	8.74	46.73	88.73	135.46
3		0.0566	4.20	1.61	6.78	9.16	46.64	100.52	147.16
4		0.0497	3.80	2.78	10.57	7.79	45.07	56.38	101.45
5		0.0400	3.50	1.59	5.58	6.02	48.50	94.65	143.15
6		0.0378	3.43	1.81	6.22	6.08	49.47	88.87	138.34
7		0.0369	3.25	2.70	8.79	4.31	45.80	43.26	89.06
8		0.0371	3.16	1.86	5.89	5.72	43.13	82.89	126.02
9		0.0413	3.20	1.63	5.22	6.82	39.71	101.31	141.02
10		0.0459	3.61	1.93	6.98	7.05	44.66	79.58	124.25
11		0.0498	3.79	1.99	7.55	7.40	44.78	74.67	119.45
12		0.0522	3.75	2.59	9.71	8.02	41.95	59.32	101.27
					Averages		**45.43**	**77.66**	**123.10**

Table 5. Comparison of DHW average energy parameters in the period of 2016–2018

Years	E_{wh}	E_{wkh}	E_a	E_a/E_{wh}
	kWh/m^3	kWh/m^3	kWh/m^3	–
2016	44.79	66.61	111.40	2.49
2017	47.32	94.00	141.32	2.99
2018	45.43	77.76	123.19	2.71
Average	**45.85**	**79.46**	**125.30**	**2.73**

2. Results of monitoring of heat energy consumption for the DHW production

Monitoring of the building's thermal energy system disclosed that dwellers of the apartment buildings have not problems with the space heating system after the renovation. Meanwhile, large amounts of energy wastage were elucidated in the DHW production system. Thermal energy overuse in this system recur year after year (see data of the monitoring in Table 2–4). The data on energy consumption for DHP production in months of 2017 are visualized and shown in the bar graph of Figure 1. The green bars show amounts of heat energy consumed for heating of cold running water up to the rated temperature 55 °C. If this hot water were consumed immediately, no more heating would be required. However, the needs for heated water depend on time and quite often they are small, especially at night (dwellers are sleeping) and during the day (mostly they work or study). As a result, water traps in the DHW pipeline and cools. The piping in the apartment building is very large, so the heat loss is very significant. Circulation pumps installed in the DHW pipeline return the cooled water back to the heat exchanger in the system, which heats the water again and again. The additional thermal energy used for the reheating and its amounts for each month are shown in the same graph in red columns. They are much larger than the green columns. This means that maintaining of rated temperature of the DHW requires significantly more energy than producing hot water (green columns). This is why the DHW is very expensive in this apartment building.

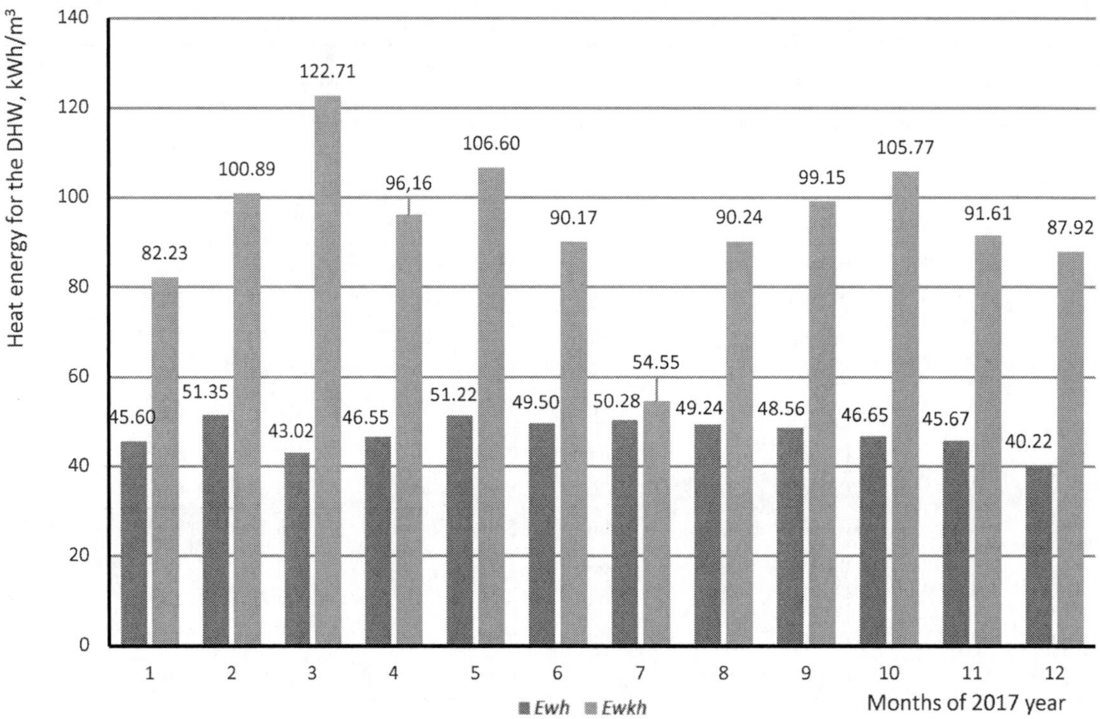

Fig. 1. Heat energy consumption for the DHW production per cubic meter of water in a typical apartment building

Figure 2 shows the heat energy amounts consumed for the DHW production per one cubic meter. The green bars show the average heat energy used for the DHW production in the years 2016, 2017, 2018 and average value of heat energy consumed during the years 2016–2018. Correspondingly, the pink bars show the average heat energy used for the DHW keeping hot in piping of the apartment building during the same years. Finally, the grey bars show the average actual (total) heat energy used for the DHW production and keeping hot in the piping of the apartment building during the same years. The graph shows that DHW system of the researched apartment building has a very high heat losses and consuming up to three times more heat than would be required. Such large overuse of thermal energy is unacceptable as it is associated with unnecessary environmental pollution and pointless waste of money.

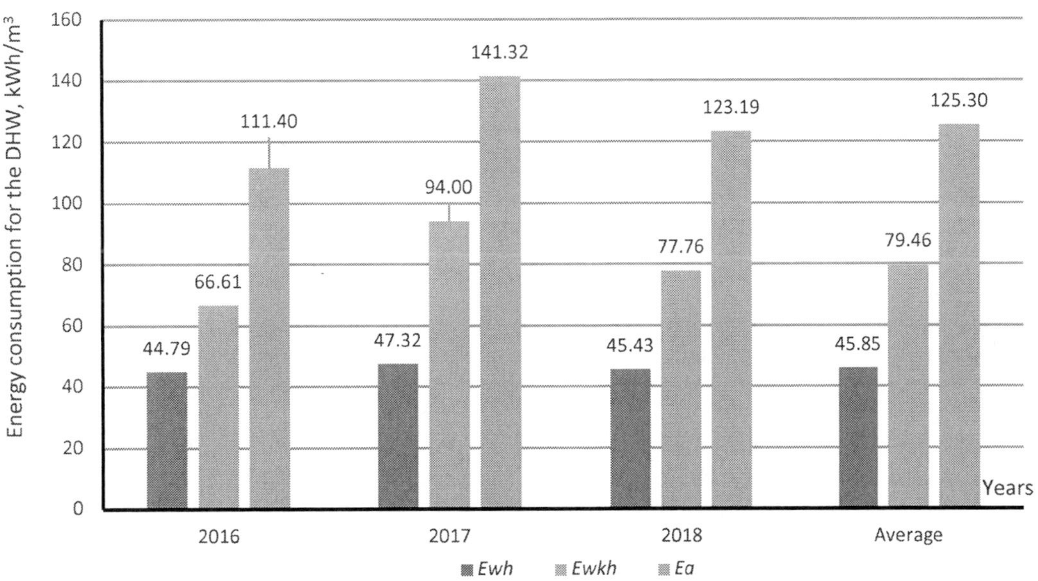

Fig. 2. Comparison of the average monthly heat energy consumption for the DHW production per cubic meter in the years 2016–2018

Studies of the space heating system have shown that the efficiency of this system meets today's requirements. Average consumption of heat energy per square meter of heated premises in this apartment building was about 40 kWh/m^2 during the heating season of 2019. The best energy efficiency indicators in this house are the apartments on the middle floors, with only one or two walls bordering the outside. Apartments on the ground floor and under the roof have slightly worse performance. Their heating efficiency is reduced by the cold coming through the floor from the basement and through the ceiling from the roof. These problems can be solved by improving the thermal insulation of the roof and basement.

Conclusions and discussion

1. Energy saving in all possible ways and fast transition to the renewable energy sources in all sectors of the global economy has become very important in the 21st century due to the rapidly accelerating global climate crisis and unprecedented levels of environmental pollution that are detrimental to all living nature and people.
2. Monitoring of the thermal energy systems of an apartment building disclosed that in the investigated building most of the thermal energy is wasted in the DHW production system.
3. Analysis of the average monthly heat energy consumption for the DHW production per cubic meter in the years 2016–2018 allowed coming to the conclusion that maintaining of rated temperature in the piping system of the DHW requires significantly more energy than producing hot water.
4. These high heat losses are the result of DHW rated temperature maintenance in the hot water pipeline and they greatly increase amounts of thermal energy E_a actually used to produce DHW and maintain its rated temperature.
5. The actual heat energy E_a is approximately 2.5 to 3 times higher than heat energy E_{wh} necessary to heat cold running water up to the rated temperature 55 °C.
6. With such a large increase in the price of DHW, it would be economically feasible to abandon such a system and use electric cold water instantaneous heaters and price of electricity 0.137 €/kWh (as presently is in Lithuania) instead – DHW produced by them would be up to 30 % cheaper.
7. The performed research of the space heating system of the apartment building allowed making sure that this system works well – average needs of heat energy of heated area were about 40 kWh/m^2 per all heating season of 2019.

However, water heating technology based on electricity is one of the most expensive. Besides, most electricity presently is generated using fossil fuels, so this technology of water heating is not very friendly to the environment. The only good feature of this technology is practically non-existent losses of thermal energy. Meanwhile, with the use of district heating, significant losses exist not only in the buildings but also along the entire route from the cogeneration plant to the building.

The simplest way to significantly improve the efficiency of the DHW system in the apartment building under study would be to create an energy-saving control program for the controller that regulates the operation of the heat exchanger (regulates temperature of hot water). This program can be created for one day or for a whole week. DHW rated temperatures could be significantly lower at night (let us say about 40 °C), when virtually all residents sleep, and during the day (when the absolute population works or studies). The hottest rated water temperature (55–60 °C) could be in the morning and after work (after learning in schools and universities) hours when DHW needs in the building are at their highest.

A few years ago, such a hot water temperature schedule was installed in this building. However, a large part of the population did not like that the DHW temperature was not always constant. They thought like this: hot water needs to be hot, and if someone needs cooler DHW water, he or she will be able to dilute it with cold water. Hot water will then be needed less and then water will cost less. Another of their "arguments" was that they pay for DHW as for a hot, but in fact quite often that water is not hot and doesn't even need to be diluted with cold water. These people simply did not realize one thing that high temperature hot water causes very large thermal energy losses in a huge hot water pipeline throughout the house despite the fact that modern thermal insulation was applied to this pipeline during the renovation of the heating unit and the building itself.

Currently there are many other mature RES-based hot water production technologies that are more cost-effective and more efficient ecologically. The following efficient hot water production technologies are already used in the buildings:

- solar collectors for water heating;
- air, ground and water sources heat pumps;
- hybrid water heating systems based on solar collectors and district heat energy supply systems [13, 14];
- hybrid water heating systems based on solar and wind power systems [11, 12] and some other.

The choice of energy sources for hot water production systems depends largely on local renewable energy sources and other circumstances. Every new house should be built with one or another renewable energy system, and old houses should be renovated in order to increase their energetic and ecological efficiency.

We believe that this article will contribute to better understanding by the dwellers of apartment buildings and other living homes owners of the processes that take place in the hot water production facilities and will encourage the implementation of the best economically and environmentally friendly solutions.

With the global movement for clean energy production and its thrift consumption, people are even being urged to switch off power to signal lights in household electrical appliances when they are not used. That appear to be extremely low power (< 1 W). However, throughout the European Union, all those indicators have a cumulative capacity equivalent to that of a nuclear power plant. Compared to this, buildings have many times greater potential for energy savings and making its production free from pollution of environment. So it is worth working hard in this area in order to contribute to the managing of the climate change process and improving human health.

References

[1] *American Lung Association. State of the Air* [Internet], 2020; 162 p. Available from: https://www.stateoftheair.org/assets/SOTA-2020.pdf (accessed on 15-08-2020).

[2] Ripple, W. J., Wolf, Ch., Newsome, T. M., Barnard, P., Moomaw, W. R. World Scientists' Warning of a Climate Emergency [Internet]. *BioScience*, 2020; 70(1):8–12. Available from: https://academic.oup.com/bioscience/article/70/1/8/5610806?searchresult=1 (accessed on 15-08-2020).

[3] Arrebola, H. J., Bullejos, M. D., Jasiūnas, K., Krawczyk, D. A., Milius, P., Rodero, S. A., Teleszewski, T. J., Urbonienė, V., Żukowski, M. *Buildings 2020+* [Internet]. Printing House of Bialystok University of Technology, 2019; 271 p. Available from: https://www.researchgate.net/publication/333699609 (accessed on 15-08-2020).

[4] Wang, R., Zhai, X. *Handbook of Energy Systems in Green Buildings*. Springer, Berlin, Heidelberg. 2018; 1885 p.

[5] *Quadrennial Technology Review. An Assessment of Energy Technologies and Research Opportunities. Chapter 5: Increasing Efficiency of Building Systems and Technologies* [Internet]. Department of Energy of the United States of America, 2015; 145 p. Available from: https://www.energy.gov/sites/prod/files/2017/03/f34/qtr-2015-chapter5.pdf (accessed on 15-08-2020).

[6] Ascione, F., Bianco, N., De Masi, R. F., Claudio De Stasio, C., Mauro, G. M., Vanoli, G. P. Multi-objective optimization of the renewable energy mix for a building [Internet]. *Applied Thermal Engineering*, 2016; 101:612–621. Available from: http://dx.doi.org/10.1016/j.applthermaleng.2015.12.073 (accessed on 01-09-2020).

[7] Sharafi, M., ElMekkawy, T. Y., Bibeau, E. L. Optimal design of hybrid renewable energy systems in buildings with low to high renewable energy ratio [Internet]. *Renewable Energy*, 2015; 83:1026–1042. Available from: http://dx.doi.org/10.1016/j.renene.2015.05.022 (accessed on 01-09-2020).

[8] Stadler, P., Ashouri, A., Maréchal, F. Model-based optimization of distributed and renewable energy systems in buildings [Internet]. *Energy and Buildings*, 2016; 120:103–113. Available from: http://dx.doi.org/10.1016/j.enbuild.2016.03.051 (accessed on 01-09-2020).

[9] Lu, Y., Wang, Sh., Zhao, Y., Yan, Ch. Renewable energy system optimization of low/zero energy buildings using single-objective and multi-objective optimization methods [Internet]. *Energy and Buildings*, 2015; 89:61–75. Available from: http://dx.doi.org/10.1016/j.enbuild.2014.12.032 (accessed on 01-09-2020).

[10] Hansen, K., Connolly, D., Lund, H., Drysdale, D., Thellufsen, J. Z. Heat roadmap Europe: identifying the balance between saving heat and supplying heat [Internet]. *Energy*, 2016; 115:1663–1671. Available from: http://dx.doi.org/10.1016/j.energy.2016.06.033 (accessed on 01-09-2020).

[11] Adomavičius, V., Valickas, J., Petrauskas, G., Pušinaitis, L. Potential of village house for sustainable energy production [Internet]. *Proceedings of 6th International Conference on Trends in Agriculture Engineering. Part I.* Prague, Czech Republic, 7–9 September, 2016; pp. 17–25. Available from: https://www.researchgate.net/publication/308305426_POTENTIAL_OF_VILLAGE_HOUSE_FOR_SUSTAINABLE_ENERGY_PRODUCTION (accessed on 05-09-2020).

[12] Adomavičius, V., Valickas, J., Petrauskas, G., Pušinaitis, L. Concept of RES-based microgrid for apartment building [Internet]. *The 11th International Conference on Intelligent Technologies in Logistics and Mechatronics Systems (ITMS'2016).* Medimond Publishing Company, 2016; pp. 1–6. Available from: https://www.researchgate.net/publication/309430727_Concept_of_RES-based_microgrid_for_apartment_building (accessed on 05-09-2020).

[13] Adomavičius, V., Petrauskas, G. Hybrid water heating system based on solar collectors and heat exchanger fed from the centralized heat supply grid [Internet]. *Proceedings of International Conference Electrical and Control Technologies.* Kaunas: KTU, 2009; pp. 352–355. Available from: https://www.researchgate.net/publication/268002128_HYBRID_WATER_HEATING_SYSTEM_BASED_ON_SOLAR_COLLECTORS_AND_HEAT_EXCHANGER_FED_FROM_THE_CENTRALISED_HEAT_SUPPLY_GRID (accessed on 05-09-2020).

[14] Adomavičius, V., Petrauskas, G. Optimisation of hybrid water heating system based on solar collectors and heat exchanger in order to maximize solar heat energy's share. *Proceedings of International Conference Electrical and Control Technologies – 2009.* Kaunas: KTU, 2009; pp. 356–361. Available from: https://www.researchgate.net/publication/278727992_OPTIMISATION_OF_HYBRID_WATER_HEATING_SYSTEM_BASED_ON_SOLAR_COLLECTORS_AND_HEAT_EXCHANGER_IN_ORDER_TO_MAXIMIZE_SOLAR_HEAT_ENERGY'S_SHARE(accessed on 05-09-2020).

Forum on Innovative Technologies and Management for Sustainability – ITMS'2020

panevėžys faculty of technologies and business

International Scientific Conference Intelligent Technologies in Logistics and Mechatronics Systems – ITELMS'2020, 1st October, 2020, Panevėžys, Lithuania

A Traffic Estimation Method for Dynamic Capacity Adaptation Targeting Energy Savings in Load Adaptive Communication Networks

Andreas Ahrens[a]*, Christoph Lange[b], Ojaras Purvinis[c], Jelena Zaščerinska[d]

[a]*Hochschule Wismar, University of Applied Sciences, Technology, Business and Design, 14 Philipp-Müller-Straße St, Wismar 23966, Germany*
[b]*Hochschule für Technik und Wirtschaft (HTW) Berlin, University of Applied Sciences, 8 Treskowallee St, Berlin 10318, Germany*
[c]*Kaunas University of Technology, Panevėžys Faculty of Technologies and Business, 33 Nemuno St, Panevėžys LT-37164, Lithuania*
[d]*Centre for Education and Innovation Research, 33 Dammes St, Riga LV-1069, Latvia*

Abstract

A challenging task for energy-efficient and sustainable network operation is the load-adaptive operation of communication networks and hence its network elements such as routers, switches and access multiplexers. Since the traffic is temporally fluctuating, load-adaptive control of the network requires a robust traffic demand estimation. The aim is to develop a simulation model for dynamic capacity adaptation based on the analysis of Wiener filtering for traffic prediction underpinning energy savings in communication network. The results show that the capacity dimensioning based on the proposed Wiener filtering traffic forecasting leads to reliable outcomes in terms of predicted traffic enabling sustainable and efficient network operation.

Keywords: energy-efficient business; sustainable network operation; network energy efficiency; load-adaptive network operation; load-adaptive control; fluctuating traffic; traffic estimation; dynamic capaticy adaptation; Wiener filtering.

Introduction

The European Union faces major challenges from the increased threats of climate change, with serious consequences in the energy sector [1]. Energy efficiency in general and particularly in the telecommunications sector, due to its tremendous and pervasive development, needs to be improved. In order to prevent dangerous climate change, the telecommunications sector is working to save energy in large-scale telecommunications networks.

Also, sustainability and reliability of modern business requires stable and efficient telecommunication networks. For example, the coronavirus pandemic (COVID-19) of 2020 as well as other unforeseen circumstances show the

* Corresponding author. Tel.: + 49 3841 753330; fax: + 49 3841 753130.
E-mail address: andreas.ahrens@hs-wismar.de

DX20F0003

importance of reliable and also economical communication network operation to maintain industry processes as far as possible by virtual means. This in turn calls for low network energy demand to meet climate change targets.

According to the latest numbers of the International Telecommunication Union (ITU), the telecommunications sector contributes around 2 % to 2.5 % of greenhouse gas emissions. Here, fixed and mobile telecommunications contribute an estimated 24 % of the total. As the ICT industry is growing faster than the rest of the economy, this share may well increase over time [2, 3].

In large-scale (nationwide) communication networks there are two drivers of energy consumption:

- coverage;
- network capacity.

Often, the first task of a network is to connect all endpoints that seek network connection. This leads to certain amount of energy demand for the active network equipment. In case the coverage is achieved, and more capacity is needed, the (excess) energy consumption is driven by the additional capacity to be installed. As today‘s network equipment is merely designed to follow the temporal fluctuating traffic demands with its provided capacity, this excess capacity is installed statically, and thus leads to a (static) additional energy demand. The present work focuses on network capacity in relation to efficiency of energy consumption.

Large-scale telecommunication networks are utilized by consumers' and business applications, such as video and music streaming but also for large file transfers for e.g. data backups. The electricity consumed by the network operation depends strongly on the installed network capacity and affects the operators 'energy bills considerably. In order to become more environment-friendly, network operators have recognized the ability for improvements throughout the recent past for reducing the network energy demand – and thus the operators' energy bill. The expected traffic amount in a network requires a certain network capacity to be installed and, therefore, determines the electricity demand of such a network via the necessary active network equipment.

A challenging and crucial task for energy-efficient and sustainable network operation is the load-adaptive operation of network elements such as routers, switches and access multiplexers. A suitable measure for roughly estimating the energy efficiency of a network system or section is the energy per bit – describing the energy needed on average for the transmission of a bit by a particular technology or network section [4]. With the technology progress to higher transmission rates, energy efficiencies can only be achieved by decreasing the energy per transmitted (processed) bit [4].

The traffic observed in networks varies between weekdays and weekends [5]. The real observed traffic depends furthermore strongly on the particular geographical and functional network section under consideration. For example, the access network in a residential area on a working day exhibits a traffic characteristic that is very different from a similar network section in a business park on the same day. However, the absolute differences between observed traffic characteristics are not the main focus of this paper. Here, we focus on a principle technique of estimating traffic patterns as an inevitable precondition for dynamic capacity provisioning that works independently of the concrete traffic curve.

A possible solution for improving the energy efficiency of networks is considered by load-adaptive operation where the network capacity follows the traffic demands. This is in contrast to the prevalent network design, where the network capacity is above the expected peak traffic plus a capacity reserve. In order to achieve any improvements in the network's energy efficiency, it is essential to adapt the provided network capacity to the fluctuating traffic demands and thus, in turn, to estimate the traffic demand reliably for these capacity dimensioning purposes. Dynamic capacity provisioning – or load-adaptive operation – can be achieved by, e.g. switching on and off ports and links to provide the necessary network capacity – or it can be done on a per–link basis. Examples for such a kind of dynamic load-adaptive network operation on a per-link basis can be found as Energy Efficient Ethernet on Ethernet links [6], they are standardized as low power mode regimes for ADSL (Asymmetric Digital Subscriber Line) connections, and, furthermore, they are discussed as radio access network management approaches [7]. Since the traffic is temporally fluctuating, load-adaptive control of the network requires a robust traffic demand estimation. This is also of overwhelming importance, as a stable network operation is a central task of network operators – since it is expected by their customers as service they pay for. Once a robust traffic prediction solution is found – that is needed for either type of load-adaptive network operation – load-adaptive network operation regimes can be considered a strong possible solution towards network energy efficiency improvements. Here, Wiener filtering has been identified as a robust solution for reliable traffic demand forecasting on relevant time scales [4].

The aim of the work is to develop a simulation model for dynamic capacity adaptation based on the analysis of Wiener filtering for traffic prediction underpinning energy savings in load adaptive network operation regimes.

It should be pointed that the term "simulation model" has been widely discussed within many scientific fields such as engineering, business and computer sciences. In order to identify the term "simulation model", the terms "simulation", "model" as well as "simulation model" are analyzed below.

Simulation means a whole environment in which a task or problem is set to which the participants react [8]. The purpose of simulation is to have participants interact in meaningful and realistic contexts [9], generating their own inter-connections.

The term "model" is of great research interests as well. In pedagogy, by model, a pattern is meant [10]. In mathematics, a model is an interpretation of a theory [11]. In engineering, business and computer sciences, a model describes a system [12]. Interdisciplinary (pedagogy, mathematics, engineering, business and computer sciences) analysis of the term "model" leads to such a newly defined notion of the term "model" as a pattern of individual's or individuals' interpretation of a phenomenon [13]. It should be noted that models can be presented in a variety of forms such as verbal, graphic, computer, etc. A model can be characterized as demonstrated in Figure 1 [13]. The model characteristics are described by parameters [13].

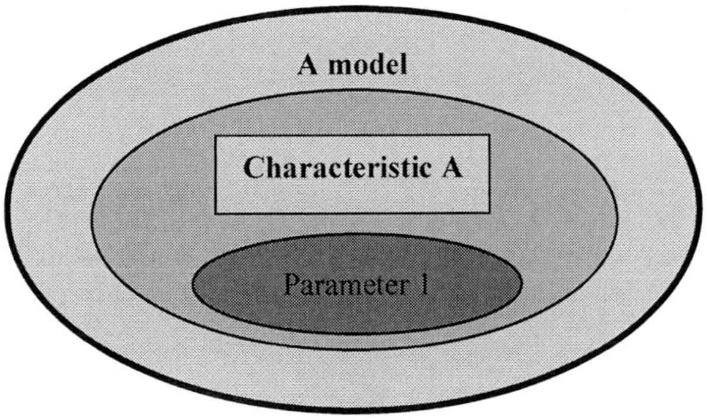

Fig. 1. Model elements [13]

By a parameter, a definable, measurable, and constant or variable characteristic, dimension, property, or value, selected from a set of data (or population) to understanding a situation (or in solving a problem) is meant [14].

In turn, a simulation model presents the behavior of a system that evolves over time [12]. Interdisciplinary (pedagogy, engineering, business and computer sciences) analysis of this definition allows a newly defined notion of the term "simulation model" as patterns of the management of phenomenon change in real-world situations [13]. A simulation model should map the characteristic of the real world process, such as load adaptive network operation regimes, with the required precision [13].

The present research employs the qualitative methodology as model creation is a qualitative process. Qualitative process is a methodology mostly used within the interpretive approach [15]. Hence, the research is carried out within the interpretive paradigm. The interpretive paradigm implies the implementation of analysis while considering a phenomenon's context. Interpretative paradigm is characterized by the researcher's practical interest in the research question [16]. The researcher is the interpreter [17].

The novelty of this contribution is the simulation model defined by the method for traffic prediction based on the Wiener filtering [18, 19, 20] as it is known from statistical signal processing: The knowledge regarding traffic behavior from the past, e.g. from previous hours or days, is used to estimate the future traffic characteristics.

The remaining part of this paper is structured as follows: in section 1, a traffic-related system model is constructed, following by the traffic prediction highlighted in section 2. In section 3, the application of traffic prediction to dynamic load-adaptive network operation is studied. The obtained results are introduced and analyzed in section "Conclusions". For verification purposes, the originally observed traffic is compared with the estimated traffic. Also, the energy consumption associated with the newly proposed capacity dimensioning strategy is calculated and compared to conventional procedures. The presented concept is verified by means of a statistical analysis where the stochastic traffic characteristics are varied, and the resulting capacity dimensioning and energy efficiency is analyzed. Concluding remarks are provided in last section.

1. Traffic-related system model

As a basis for establishing traffic prediction algorithms, real measured traffic data or a modeled traffic time function with suitable characteristics and statistics are necessary. Throughout this paper a traffic model is used that refers to an exemplary link in a network whose capacity is subject to load-adaptive switching regimes. The traffic function is constructed as follows: An underlying time function *s(k)* (e.g. mean traffic), with variations on a longer time scale, is used for modelling the average traffic fluctuation observed for an exemplary link throughout a day. To model the stochastic variations in the traffic on a shorter time scale, an additive white Gaussian noise *n(k)* with zero mean and the variance P_R is added. In consequence, the traffic function *v(k)* is obtained that is referred to an observed traffic throughout the paper. The observed (measured) traffic *v(k)* results in:

$$v(k) = s(k) + n(k). \tag{1}$$

Figure 2 shows exemplary curves of the observed (measured) traffic *v(k)* and the underlying averaged traffic function *s(k)*.

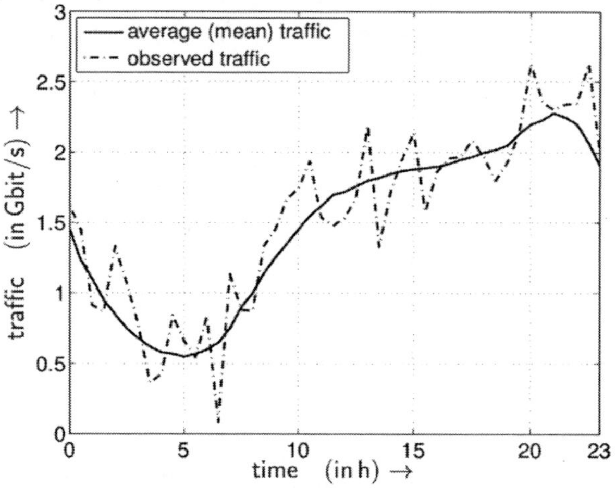

Fig. 2. Characteristics of exemplarily averaged (solid line) and observed traffic (dashed line)

The resulting system model is highlighted in Figure 3.

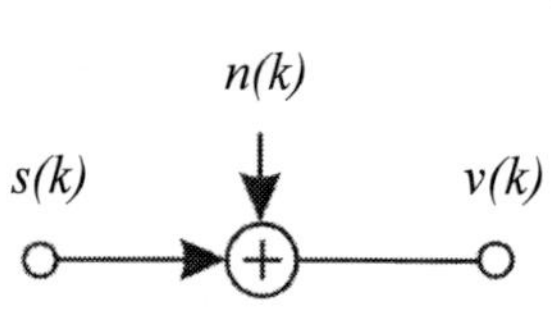

Fig. 3. Resulting system model for modelling traffic fluctuations

This modelled traffic contains the long-term traffic fluctuations over a day as well as the inherent stochastic nature of typical broadband data traffic. The complementary cumulative distribution function (CCDF) of the modelled traffic at noon is shown in Figure 4.

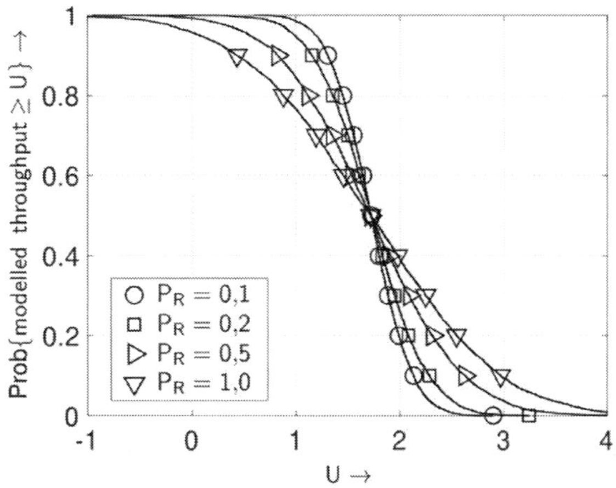

Fig. 4. CCDF of the throughput at noon taking different values of P_R into account

Assuming a throughput of 1.72 Gbit/s (averaged traffic observed *s(k)*) at noon traffic fluctuations become obvious. In conclusion, in this way an appropriate traffic model has been obtained that can be described and adjusted by analytic parameters. In Figure 4 exemplarily variances PR, describing the short-term fluctuations, in the range of 0.1…1.0 are selected. The noise is used to simulate the random fluctuations of the traffic in a network on a very short time scale that cannot be described by a deterministic function. The variance of that noise is chosen in a way that realistic orders of magnitude are met – that can be observed on real network links. The variance of a bit rate – that is measured in bit/s – exhibits the dimension of a bit rate squared, i.e. the variance describing the short-term traffic fluctuations has technically the unit (bit/s). In the interest of the clarity of the presentation in this work, the units of this variance are omitted.

2. Traffic prediction using Wiener filtering

For a robust traffic estimation, a Wiener filter is used in this work, since it is suitable for tasks when minimizing the mean square error (MMSE – *Minimum Mean Square Error*) between the estimated (i.e. predicted) traffic and the real traffic. The Wiener filtering approach is in particular viable, when the mean traffic is affected by short-term fluctuations that are modelled as Gaussian noise. Therefore, differences in the mean traffic such as between weekdays and weekends are not taken into consideration as these fluctuations are considered by the mean value – that differs between weekdays and weekend days and thus leads to different numerical values and results, but has no impact on the considered approach. A linear predictor can be used to estimate the traffic at the time k by taking the last q traffic values $v[k-q]$ into account and results in

$$\hat{v}[k] = \sum_{\mu=1}^{q} p_\mu \cdot v[k-\mu]; \quad (2)$$

with the parameter q describing the order of the predictor. The coefficients of the predictor $p\mu$ (for $\mu = 1, 2, \ldots, q$) have to be defined by minimizing the energy of the error signal $e[k] = v[k] - \hat{v}[k]$. The error signal $e(k)$ appears after linear filtering of the signal $v(k)$ with the so far unknown filter coefficients $b(k)$ (Fig. 5) – which are related to the predictor coefficients by $b[\mu] = -p[\mu]$ for $1 \leq \mu \leq q$, $b[0] = 1$ and $b[\mu] = 0$ for all other μ. Details on the derivation of this interrelationship are shown in [21].

Taking the stationary mean (averaged) traffic $s(k)$ and the added noise $n(k)$ into account, the observed noisy process $v(k)$ forms the basis for the proposed traffic prediction. Using the Wiener filter the mean square error between the estimated traffic $\hat{v}(k)$ and the mean (averaged) traffic $s(k)$ can be minimized.

In Figure 6 the curves of the exemplary observed traffic $v(k)$ and the predicted traffic $\hat{v}(k)$ are shown: It becomes obvious that the estimated or predicted time function follows the observed traffic in tendency but is not directly useful for capacity dimensioning – as there are time periods where the traffic is under-estimated.

Therefore, some modification or adaption of the Wiener filtering is necessary for capacity dimensioning purposes in order to take those deviations into account.

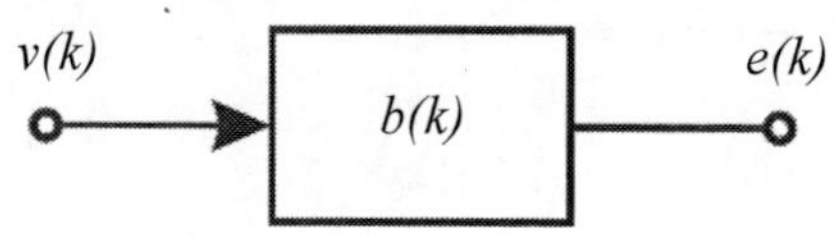

Fig. 5. Error signal $e(k)$ as a function of traffic function $v(k)$

The target is always a reliable network operation – meaning here sufficient capacity – and then somewhat downstream – the improved energy efficiency.

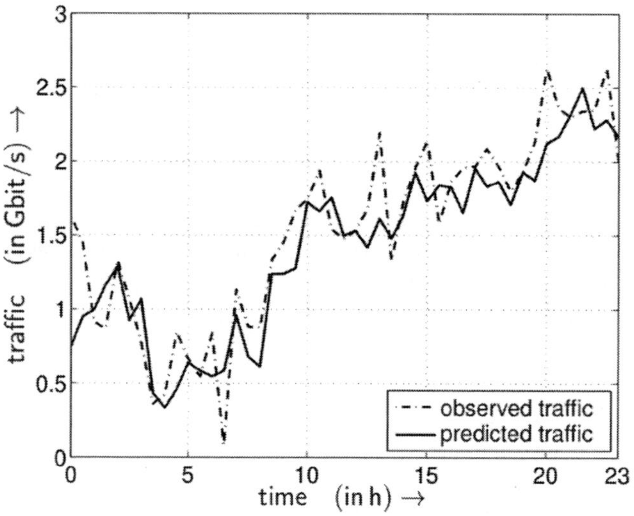

Fig. 6. Observed (dashed line) and predicted (solid line) traffic time functions

The target is always a reliable network operation – meaning here sufficient capacity – and then somewhat downstream – the improved energy efficiency.

3. Capacity dimensioning and energy efficiency

Provided that there is a linear dependency between capacity and power ensured by the network elements, from the capacity time function $c(t)$ a power time function $P(t)$ is obtained by $P(t) = K \cdot c(t)$, where the factor K exhibits the dimension of an energy per bit (in J/bit or Ws/bit). The actual value and magnitude of K depends strongly on the system technologies and their generations. The energy consumed by a bit of data as it runs through a telecommunication network, e.g. the Internet, can be estimated by counting the number of network elements – e.g. switches, routers, amplifiers, transceivers – that the bit passes through, and adding all of these contributions to the energy consumption of that bit of data. According to [22] it is expected that a high-end core router consumes around 20 nJ/bit, while Ethernet switches consume less than 10 nJ/bit. These numbers depend strongly on the technologies and therefore are subject to improve as technology improves.

In this work the parameter K is assumed to be $K = 10^{-6}$ Ws/bit. The value of energy per bit is determined by the communication system in use e.g. a switch, a router or an access multiplexer – or network section under consideration e.g. optical access network, core network or the radio link. A thorough investigation on this topic with typical numerical values can be found e.g. in [22]. The chosen value of $K = 10^{-6}$ Ws/bit is a typical value out of a wide range of possible values.

Taking into account that the power consumption function $P(t)$ follows the traffic function $v(t)$, $P(t)$ has to be adapted according to the traffic (Fig. 7). As highlighted by Figure 7 load adaptiveness leads to energy efficiency improvements.

Now temporal power consumption $P(t)$ is no longer constant. To measure energy efficiency improvement of particular load-adaptive case n, energy efficiency parameter $\varepsilon_n = E_n / E_0$ is used, as defined in [23, 24]. Here, $E_0 = P_0 \cdot T$ describes the reference case with no load-adaptiveness at all.

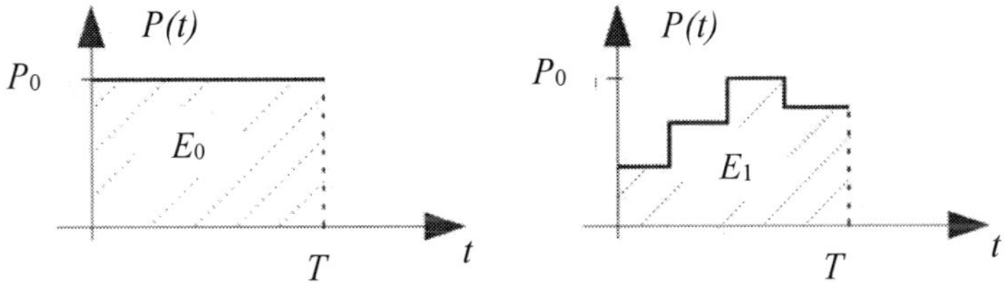

Fig. 7. Energy efficiency improvement by taking power traffic-dependent time function for load-adaptiveness (right) and non load-adaptiveness (left) into account

4. Energy efficiency results

Based on the capacity dimensioning use cases in Figure 8, the capacity follows directly the estimated traffic. As an example a noise variance $P_R = 0.1$ is assumed for describing the short-term traffic fluctuations. In order to avoid a capacity bottleneck, a traffic reserve Δ is added to the traffic $\hat{v}(t)$, i.e. $c(t) = \hat{v}(t) + \Delta$, to ensure a sufficient capacity. This traffic reserve Δ is especially needed for situations where the real traffic is under-estimated by the predictor. The energy efficiency of different cases of load-adaptive operation regimes is shown in Figure 9.

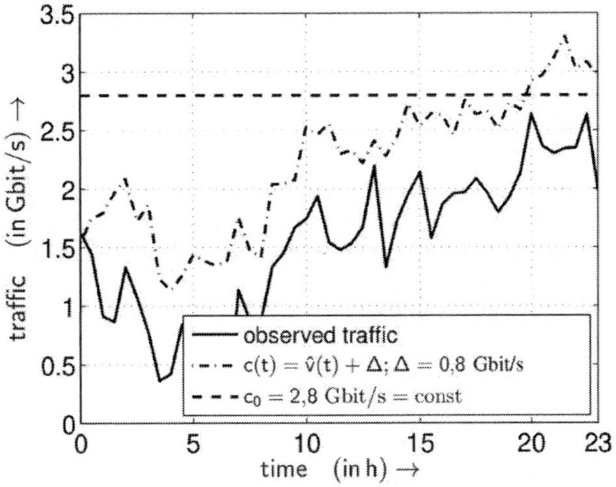

Fig. 8. Capacity as a function of the estimated traffic for different parameters of the traffic reserve

Hereby, scenario 0 describes the reference case employing no load-adaptiveness at all and scenario 1 represents the best-case limit, where the capacity follows the observed traffic ideally. Realistic load-adaptive regimes will exhibit energy efficiencies ε_n between those boundaries. It becomes obvious that energy efficiency is increased when approximating the traffic curve more exactly.

However, in scenarios where the traffic is under-estimated a capacity bottleneck could appear.

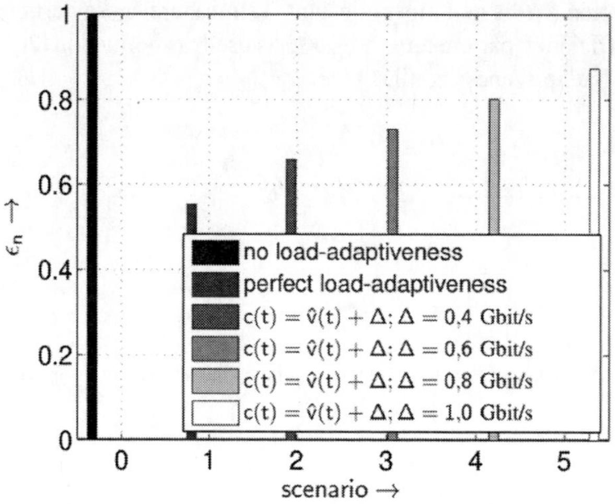

Fig. 9. Energy efficiency for different parameters of the traffic reserve based on $c_0 = 2.8$ Gbit/s

The probability will doubtlessly increase for lower Δ. Therefore, the parameter Δ has to be selected carefully.

Conclusions

In this paper, a traffic prediction approach for temporally fluctuating network traffic based on Wiener filtering has been analysed. The results show that the capacity dimensioning based on the proposed Wiener filtering traffic forecasting leads to reliable outcomes in terms of predicted traffic enabling sustainable and efficient network operation.

The findings of the theoretical analysis allow creating the simulation model of dynamic capacity adaptation based on the analysis of Wiener filtering for traffic prediction underpinning energy savings load adaptive network operation regimes.

Our approach can be useful in case the capacity will be provided basing on the temporally fluctuating traffic demands. If the excess capacity is provided as usual by statically adding more lines for additional capacity only the fact whether a port of a line is on or not will determine the energy consumption. Before the background of increased energy cost and increasing sensibility for environmental concerns capacity in the future should be provided load-adaptively – and then traffic prediction algorithms like discussed in this paper are indispensable and come in handy for network design and planning.

The presented work is limited by the creation of the simulation model only. Another limitation is the application of Wiener filtering for prediction of energy consumption on a communication network component or larger site based on past traffic capacity.

Future work will focus on validation of the proposed simulation model. Further on, validation of the simulation model will be implemented in different environments. Enhancement of the "model" definition, the present work refers to, is also planned. The use of other approaches and paradigms, in comparison to the interpretive paradigm, for analysing further research results will be included in future work. Analysis of other prediction methods will be carried out, too. A comparative study of different prediction methods will be presented. Deep analysis of the interrelations between energy and traffic as well as network port will be implemented. Modification of network capacity based on the load will be analysed. Treatment of traffic volumes on different days, e.g. weekday vs. weekend, will be detailed.

References

[1] Amanatidis, G. *European Policies on Climate and Energy Towards 2020, 2030 and 2050.* Policy Department for Economic, Scientific and Quality of Life Policies Directorate-General for Internal Policies, 2019. PE 631.047.

[2] Roy, S. N. Energy logic: a road map to reducing energy consumption in telecommunications networks. *International Telecommunications Energy Conference*, San Diego (USA), 2008; pp. 4–2,

[3] Marsch, P., Bulakci, O., Queseth, O., Boldi, M. *5G System Design: Architectural and Functional Considerations and Long Term Research.* John Wiley & Sons, Chichester, 2018.

[4] Ahrens, A., Lange, C., Zaščerinska, J. Energy savings by using traffic estimation for dynamic capacity adaptation in communication network operations. *Journal of Communications,* 2020; 15(11):790–795.

[5] Betker, A., Gamrath, I., Kosiankowski, D., Lange, C., Lehmann, H., Pfeuffer, F., Simon, F., Werner, A. Comprehensive topology and traffic model of a nationwide telecommunication network. *Journal of Optical Communications and Networking*, (2014); 6(11):1038–1047.

[6] Reviriego, P., Christensen, K., Rabanillo, J., Maestro, J. A. An initial evaluation of energy efficient Ethernet. *IEEE Communications Letters*, 2011; 15(5):578–580.

[7] Ambrosy, A., Blume, O., Klessig, H., Wajda, W. Energy saving potential of integrated hardware and resource management solutions for wireless base stations. *22nd International Symposium on Personal Indoor and Mobile Radio Communications* (PIMRC), Toronto (Canada), 2011.

[8] Stutridge, G. Using simulations in teaching English for specific purposes. *Modern English Publications* Ltd. Printed by Linneys of Mansfield, 1977; pp. 32–34

[9] Porto, B. C. Developing speaking skills by creating our own simulations for the EFL courses. *English Language Teaching*, 2019; 19(2):3–20.

[10] Belickis, I., Bluma, D., Koke, T., Markus, D., Skujina, V., Šalme, A. *Pedagoģijas Terminu Skaidrojošā Vārdnīca* [Dictionary of pedagogical terminology]. Rīga: Zvaigzne ABC, 2000; 248 p.

[11] Kühne, T. What is a model? *Dagstuhl Seminar Proceeding*. Publisher: Internationales Begegnungs und Forschungszentrum für Informatik (IBFI), Schloss Dagstuhl, Germany, 2005; pp. 1–10.

[12] Banks, J., Carson, J., Nelson, B. L., Nicol, D. *Discrete-Event System Simulation*, 5/E. Prentice Hall Paper, 2004; 624 p.

[13] Ahrens, A., Purvinis, O., Zaščerinska, J., Andreeva, N. Quasi-group decision making in higher education: a model for analysis of binary students' behaviour. *Conference: III International Scientific Practical Conference "Trends in Science and Studies Under Conditions of Globalisation"*, Panevėžys College, Panevėžys, Lithuania, 2015; 1:45–56. ISSN 2029-1280.

[14] Business Dictionary. *Parameter.* BusinessDictionary.com, 2015.

[15] Thanh, N. C., Thanh, T. T. L. The interconnection between interpretivist paradigm and qualitative methods in education. *American Journal of Educational Science*, 2015; 1(2):24–27.

[16] Cohen, L., Manion, L., Morrsion, K. *Research Methods in Education* (5th ed.). London and New York: Routledge / Falmer Taylor and Francis Group, 2005.

[17] Ahrens, A., Purvinis, O., Zaščerinska, J., Miceviciene, D., Tautkus, A. *Burstiness Management for Smart, Sustainable and Inclusive Growth: Emerging Research and Opportunities.* IGI Global, Hershey, 2018.

[18] Wiener, N. *Extrapolation, Interpolation, and Smoothing of Stationary Time Series.* Wiley, New York, 1949.

[19] Vega, L. R., Rey, H. *A Rapid Introduction to Adaptive Filtering.* Springer, Heidelberg, New York, 2013.

[20] Vaseghi, S. V. *Advanced Digital Signal Processing and Noise Reduction.* John Wiley & Sons, Chichester, 2009.

[21] Ahrens, A., Lange, C., Benavente-Peces, C. Traffic estimation for dynamic capacity adaptation in load adaptive network operation regimes. *International Conference on Pervasive and Embedded Computing and Communication Systems (PECCS*), Lisbon (Portugal), 2016.

[22] Tucker, R. S. Green optical communications part II: energy limitations in networks. *IEEE Journal of Selected Topics in Quantum Electronics*, 2011; 17(2):261–274.

[23] Lange, C., Gladisch, A. Limits of energy efficiency improvements by load-adaptive telecommunication network operation. *10th Conference of Telecommunication, Media and Internet Techno-Economics* (CTTE), Berlin (Germany), 2011; pp. 5–1.

[24] Lange, C., Kosiankowski, D., Betker, A., Simon, H., Bayer, N., Hugo, D., Lehmann, H., Gladisch, A. Energy efficiency of load-adaptively operated telecommunication networks. *IEEE/OSA Journal of Lightwave Technology*, 2014; 32(4):571–590.

Forum on Innovative Technologies and Management for Sustainability – ITMS'2020

ktu 1922 panevėžys faculty of technologies and business

International Scientific Conference Intelligent Technologies in Logistics and Mechatronics Systems – ITELMS'2020, 1st October, 2020, Panevėžys, Lithuania

Multi-Objective Assessment of the Effectiveness of the Thermal Insulation Layer of the Flat Roof by EDAS Method

Donatas Aviža[a, b]*, Danguolė Striukienė[a], Elvyra Zacharovienė[a]

[a]*Kaunas University of Technology, Panevėžys Faculty of Technologies and Business, 33 Nemuno St, Panevėžys LT-37164, Lithuania*
[b]*Panevėžys University of Applied Sciences, Faculty of Technological Sciences, 23 Laisvės Sq., Panevėžys LT-35200, Lithuania*

Abstract

In the course of the empirical research, the efficiency of the middle thermal insulation layer of the A++ class flat roof of a public building on corrugated steel supporting structure was assessed. The second fire resistance class was adopted. EDAS method was selected for the multi-objective assessment of the effectiveness of the roof thermal insulation alternatives. Four indicators were evaluated: the price of the thermal insulation; the design value of the thermal conductivity coefficient; the density and the thickness of the layer. Findings of the performed empirical research suggest that the polystyrene foam *Šiloporas Neo EPS 100 N* is the most effective thermal insulation for the flat roof. This alternative scored the highest number of points (1 point of 1). The least effective option (from the tested alternatives) is the rock wool – *Paroc ROL 30*. The research findings are relevant to the construction structural designers and investors.

Peer-review under responsibility of the Kaunas University of Technology, Panevėžys Faculty of Technologies and Business.

Keywords: EDAS method; roof; thermal insulation; multi-objective evaluation.

Introduction

Buildings account for 40 % of total energy consumption in the European Union. The sector is expanding, which is bound to increase its energy consumption. Therefore, reduction of energy consumption and the use of energy from renewable sources in the buildings sector constitute important measures needed to reduce the Union's energy dependency and greenhouse gas emissions. Reduced energy consumption and an increased use of energy from renewable sources also have an important part to play in promoting the security of energy supply, technological developments and in creating opportunities for employment and regional development, in particular in rural areas.

* Corresponding author. Tel.: + 370 623 75003.
E-mail address: donatas.aviza@ktu.lt

DX20F0004

Buildings have an impact on the energy consumption over the long term. Given that the renovation cycle of existing buildings is long, new buildings and the existing ones, which are under major renovation, should therefore meet the minimum energy performance requirements adopted to local climatic conditions [1].

Objectives of the research:

1. To perform the analysis of scientific resources revealing the requirements for the roofs of A++ class public buildings.
2. To analyze the EDAS method of multi-objective evaluation.
3. To perform a multi-objective evaluation study of the effectiveness of thermal insulation materials.

Research methods: the analysis of scientific and technical sources, the EDAS method and the empirical research.

According to the Lithuanian's normative requirements, buildings are classified into 9 classes according to the energy efficiency: A++, A+, A, B, C, D, E, F and G. The highest energy efficiency class is the A++ class, which is assigned to a building that consumes almost no energy [2]. The Directive 2010/31/EU of the European Parliament and of the Council of 19 May 2010 on the energy performance of buildings states that only such buildings will have to be built from 2021 onwards [1].

The nearly zero-energy buildings are the buildings, which comply with the requirements set out for A ++ energy efficiency class buildings in the Construction Technical Regulation STR 2.01.02:2016 "Design and Certification of Energy Performance of Buildings". These are highly energy efficient buildings, which consume almost zero or very low amount of energy. The renewable energy, which is mostly produced locally or come from nearby sources, accounts for the majority of energy consumption in these buildings [2].

The most of energy is lost through the roof (30–35 %), windows (21–35 %), walls (18–25 %), floor (12–14 %) and foundation (6–9 %). The biggest heat loss is through the roof, therefore the roofs are subject to the strictest requirements [3].

The issue of energy saving is relevant both at the national and international level. There is a number of studies in the scientific literature about zero-energy buildings (Lu [19]; Marszal et al. [20]; Gatt et al. [21]). Lithuanian researchers Aviža et al. [17]; Levinckytė et al. [16]; Kaupienė et. al. [18] also examine the impact of the heat loss on the "A" class energy efficiency buildings.

1. Evaluation based on distance from average solution (EDAS) method

The EDAS method uses an average solution for appraising the alternatives, considering the positive distance from the average, and the negative distance from the average. This method is very useful when conflicting criteria have to be considered. As has been found and claimed by the authors of the method [4, 6], the EDAS method is stable when various criteria weights are used, and it is consistent with other methods. In addition, the advantage of the proposed method is the simplicity and faster computation, especially as these advantages do not affect the accuracy of calculation. The steps for using the proposed method are presented as follows [5, 7, 8].

Step 1. Construction of the decision-making matrix (*X*) [9, 10]. The method evaluates the decision matrix *X*, which refers to n alternatives and m criteria (1):

$$X = \left[X_{ij}\right]_{n\times m} = \begin{bmatrix} X_{11} & X_{12} & \cdots & X_{1m} \\ X_{21} & X_{22} & \cdots & X_{2m} \\ \vdots & \vdots & \ddots & \vdots \\ X_{n1} & X_{n2} & \cdots & X_{nm} \end{bmatrix}; \quad (1)$$

where X_{ij} is the performance of the *i*-th alternative with regard to the *j*-th criterion.

Determination of average solution (AV_j) (2) according to all criteria [11, 13, 14]:

$$AV_j = \frac{\sum_{i=1}^{n} X_{ij}}{n}. \quad (2)$$

Step 2. The positive distance from average (*PDA*) is calculated according to the criteria, which are beneficial or non-beneficial:

- if j-th criterion is beneficial (3):

$$PDA_{ij} = \frac{max\left(0,(X_{ij}-AV_j)\right)}{AV_j}; \tag{3}$$

- if j-th criterion is non-beneficial (4):

$$PDA_{ij} = \frac{max\left(0,(AV_j-X_{ij})\right)}{AV_j}. \tag{4}$$

Step 3. Calculation of positive distance from the average (*PDA*) and negative distance from the average (*NDA*) [15]:

- if j-th criterion is beneficial (5):

$$NDA_{ij} = \frac{max\left(0,(AV_j-X_{ij})\right)}{AV_j}; \tag{5}$$

- if j-th criterion is non-beneficial (6):

$$NDA_{ij} = \frac{max\left(0,(X_{ij}-AV_j)\right)}{AV_j}. \tag{6}$$

Step 4. The Weighted sum of *PDA* is obtained from the average matrix (7):

$$SP_i = \sum_{j=1}^{m} w_j PDA_{ij}. \tag{7}$$

Step 5. The Weighted sum of *NDA* is obtained from the average matrix (8):

$$SN_i = \sum_{j=1}^{m} w_j NDA_{ij}. \tag{8}$$

Step 6. Calculate normalize the values of *SP* and *SN* (9, 10):

$$NSP_i = \frac{SP_i}{max_i(SP_i)}; \tag{9}$$

$$NSN_i = \frac{SN_i}{max_i(SN_i)}. \tag{10}$$

Step 7. Normalize the values of *NSP* and *NSN* (11) [15]:

$$AS_i = \frac{1}{2}(NSP_i + NSN_i). \tag{11}$$

Step 8. According to the obtained AS_i, alternatives are ranked in descending order. The alternative with the highest *AS* is the best one among the other alternatives [5].

2. Research

The research assessed the efficiency of the middle thermal insulation layer of the A++ class flat roof of a public building on the corrugated steel supporting structure. Heat transfer coefficient – $U = 0.11$ W/m²K. The buildings were considered to be of fire resistance class 2. The research model is presented in Figure 1. The EDAS method was chosen for the multi-objective evaluation of the effectiveness of the thermal insulation alternatives.

In the first stage, a decision-making matrix (Table 1) is compiled and 5 different thermal insulation materials are assessed according to 4 indicators: the price of the thermal insulation; the design value of the thermal conductivity coefficient; the density; the layer thickness.

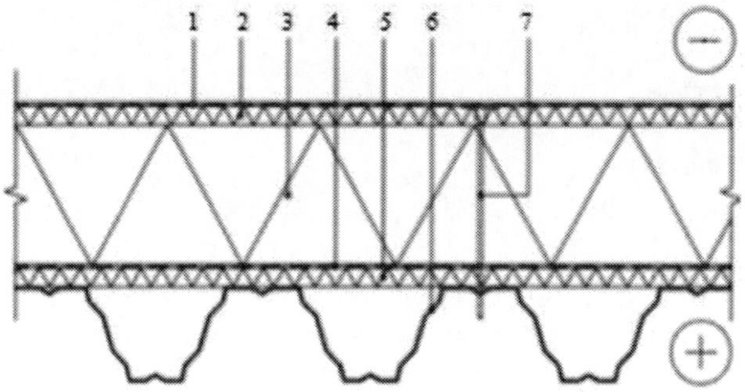

Fig. 1. Research model [22]: 1 – roof membrane; 2 – *Paroc ROB* 60; 3 – tested layer (5 materials); 4 – vapour barrier; 5 – *Paroc ROB 60*; 6 – supporting structure (corrugated steel); 7 – mechanical fasteners

The significance of all indicators is assumed to be proportionate and equal. The average values of the indicators are determined.

Table 1. Decision-making matrix

Options (different thermal insulation materials)	Indicators			
	The price of the thermal insulation, €/m^3	Design value of the thermal conductivity coefficient, W/mK	The density, kg/m^3	The layer thickness, mm
Paroc ROL 30	78.65	0.040	80.00	300.00
EPS100	54.45	0.037	18.50	280.00
XPS TECHNONICOL CARBON PROF	90.45	0.034	38.00	250.00
Šiloporas Neo EPS 100 N	65.00	0.032	18.50	240.00
XPS FINNFOAM FL-200	96.70	0.035	27.00	260.00
Average	77.05	0.036	36.40	266.00
Significance	0.25	0.250	0.25	0.25

In the second stage, the positive distance from the average (*PDA*) of the decision is determined (Table 2).

Table 2. Positive distance from average (*PDA*)

Options	Indicators			
	The price of the thermal insulation, €/m^3	Design value of the thermal conductivity coefficient, W/mK	The density, kg/m^3	The layer thickness, mm
Paroc ROL 30	0.000	0.000	0.000	0.000
EPS100	0.293	0.000	0.492	0.000
XPS TECHNONICOL CARBON PROF	0.000	0.045	0.000	0.060
Šiloporas Neo EPS 100 N	0.156	0.101	0.492	0.098
XPS FINNFOAM FL-200	0.000	0.017	0.258	0.023
Significance	0.250	0.250	0.250	0.250

In the third stage, the negative distance from the average (*NDA*) of the decision is determined (Table 3).

Table 3. Negative distance from average (*NDA*)

Options	Indicators			
	The price of the thermal insulation, €/m^3	Design value of the thermal conductivity coefficient, W/mK	The density, kg/m^3	The layer thickness, mm
Paroc ROL 30	0.021	0.124	1.198	0.128

EPS100	0.000	0.039	0.000	0.053
XPS TECHNONICOL CARBON PROF	0.174	0.000	0.044	0.000
Šiloporas Neo EPS 100 N	0.000	0.000	0.000	0.000
XPS FINNFOAM FL-200	0.255	0.000	0.000	0.000
Significance	0.250	0.250	0.250	0.250

In the fourth stage, the weighted alternative (*PDA*) values SP_i are calculated (Table 4):

Table 4. Weighted sum of *PDA*

Options	Indicators				SP_i
	The price of the thermal insulation, €/m³	Design value of the thermal conductivity coefficient, W/mK	The density, kg/m³	The layer thickness, mm	
Paroc ROL 30	0.000	0.000	0.000	0.000	0.000
EPS100	0.073	0.000	0.123	0.000	0.196
XPS TECHNONICOL CARBON PROF	0.000	0.011	0.000	0.015	0.026
Šiloporas Neo EPS 100 N	0.039	0.025	0.123	0.024	0.212
XPS FINNFOAM FL-200	0.000	0.004	0.065	0.006	0.074
Significance	0.000	0.000	0.000	0.000	0.000

In the fifth stage, the weighted alternative (*NDA*) values SN_i are calculated (Table 5):

Table 5. Weighted sum of *NDA*

Options	Indicators				SN_i
	The price of the thermal insulation, €/m³	Design value of the thermal conductivity coefficient, W/mK	The density, kg/m³	The layer thickness, mm	
Paroc ROL 30	0.005	0.031	0.299	0.032	0.367
EPS100	0.000	0.010	0.000	0.013	0.023
XPS TECHNONICOL CARBON PROF	0.043	0.000	0.011	0.000	0.054
Šiloporas Neo EPS 100 N	0.000	0.000	0.000	0.000	0.000
XPS FINNFOAM FL-200	0.064	0.000	0.000	0.000	0.064
Significance	0.005	0.031	0.299	0.032	0.367

In the sixth and seventh stages, the normalized values of SP_i and SN_i and the averages of these values AS_i are calculated, and then the alternative priority row is determined (Table 6).

Table 6. Ranking alternatives

Options	Values				AS_i	Rank
	SP_i	SN_i	NSP_i	NSN_i		
Paroc ROL 30	0.000	0.367	0.000	0.000	0.00	**5**
EPS100	0.196	0.023	0.927	0.937	0.93	**2**
XPS TECHNONICOL CARBON PROF	0.026	0.054	0.124	0.852	0.49	**4**
Šiloporas Neo EPS 100 N	0.212	0.000	1.000	1.000	1.00	**1**
XPS FINNFOAM FL-200	0.074	0.064	0.351	0.827	0.59	**3**
Max	0.212	0.367				

Upon completion of the multi-objective assessment, effectiveness alternatives (Step 8) were calculated for each thermal insulation materials (Fig. 2).

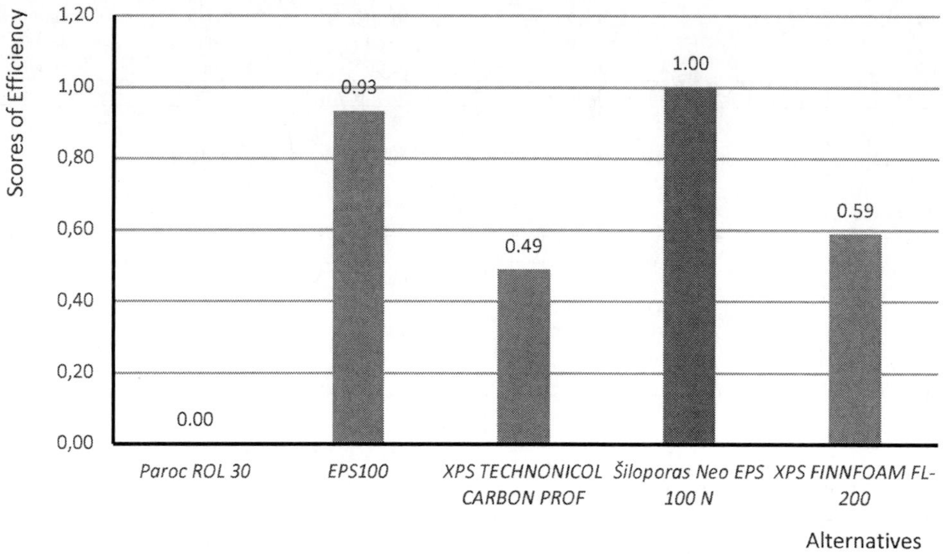

Fig 2. Evaluation of thermal insulation layer efficiency

An empirical study shows that the most effective thermal insulation layer for a public A++ class flat roof is the polystyrene foam *Šiloporas Neo EPS 100N* (1 point of 1), the least effective is rock wool *Paroc ROL 30* (0 points of 1).

Conclusions

1. Having evaluated (the price of the thermal insulation; the design value of the thermal conductivity coefficient; the density and the layer thickness of the material) by the EDAS method, the most effective thermal insulation of the middle layer of a flat roof of A++ class public building on the corrugated steel supporting structure (taking into account five alternatives) is the polystyrene foam – *Šiloporas Neo EPS 100 N*. This alternative scored the highest number of points (1 point of 1).
2. The polystyrene foam *EPS100* lagged behind and took second place, this material scored 0.93 of 1 possible and is 7 percent less effective than *Neo EPS 100 N*. However this material is the cheapest of all the options studied.
3. In the third place – the extruded polystyrene foam *XPS FINNFOAM FL-200*. This material scored 0.59 points of 1 possible and it is 41 percent less effective than the 1 place alternative.

References

[1] Directive 2010/31/EU of the European Parliament and of the Council of 19 May 2010 on the energy performance of buildings [Internet]. *OJ L* 153, 18.6.2010; p. 13–35. Available from: http://data.europa.eu/eli/dir/2010/31/oj (accessed on 01-09-2020).

[2] *STR 2.01.02:2016 Pastatų energinio naudingumo projektavimas ir sertifikavimas*. Vilnius, 2016.

[3] *STR 2.05.02:2008 Statinių konstrukcijos. Stogai*. Vilnius, 2008.

[4] Han, L., Cuiping, W. An extended EDAS method for multicriteria decision-making based on multivalued neutrosophic sets [Internet]. *Complexity Journal*, Hindawi, 2020; 2020(7578507):1–9. Available from: https://doi.org/10.1155/2020/7578507 (accessed on 01-09-2020).

[5] Kundakcı, N. An integrated method using MACBETH and EDAS methods for evaluating steam boiler alternatives [Internet]. *Journal of Multi-Criteria Decision Analysis*, 2019; 26:27–34. Available from: https://doi.org/10.1002/mcda.1656 (accessed on 01-09-2020).

[6] Keshavarz-Ghorabaee, M., Amiri, M., Zavadskas, E. K., Turskis, Z., Antucheviciene, J. A comparative analysis of the rank reversal phenomenon in the EDAS and TOPSIS methods. *Economic Computation and Economic Cybernetics Studies and Research / Academy of Economic Studies*, 2018; 52:121–134.

[7] Stanujkic, D., Zavadskas, E. K., Keshavarz Ghorabaee, M., Turskis, Z. An extension of the EDAS method based on the use of interval grey numbers. *Studies in Informatics and Control*, 2017; 26(1):5–12.

[8] Trinkūnienė, E., Podvezko, V., Zavadskas, E. K., Jokšienė, I., Vinogradova I., Trinkūnas, V. Evaluation of quality assurance in contractor contracts by multi-attribute decision-making methods. *Economic Research-Ekonomska Istraživanja*, 2017; 30(1):1152–1180.

[9] Turskis, Z., Morkunaite, Z., Kutut, V. A hybrid multiple criteria evaluation method of ranking of cultural heritage structures for renovation projects [Internet]. *International Journal of Strategic Property Management*, 2017; 21(3):318–329. Available from: https://doi.org/10.3846/1648715X.2017.1325782 (accessed on 01-09-2020).

[10] Amiri, M., Keshavarz Ghorabaee, M., Turskis, Z., Zavadskas, E. K. Multi-criteria group decision-making using an extended EDAS method with interval type-2 Fuzzy sets. *Ekonomie a Management*, 2017; 20(1):48–68.

[11] Cevik Onar, S., Kahraman, C., Keshavarz-Ghorabaee, M., Oztaysi, B., Yazdani, M., Zavadskas, E. K. Intuitionistic fuzzy EDAS method: an application to olidwaste disposal site selection. *Journal of Environmental Engineering and Landscape Management*, 2017; 25(1):1–1.

[12] Amiri, M., Keshavarz-Ghorabaee, M., Turskis, Z., Zavadskas, E. K. Extended EDAS method for fuzzy multi-criteria decision-making an application to supplier selection. *International Journal of Computers Communications &Control*, 2016; 11(3):358–371.

[13] Barauskas, A., Jakovlevas-Mateckis, K., Palevičius, V, Antuchevičienė, J. Ranking conceptual locations for a park-and-ride parking lot using EDAS method. *GRAĐEVINAR*, 2018; 11:975–983.

[14] Keshavarz-Ghorabaee, M., Zavadskas, E. K., Olfat, L., Turskis, Z. Multi-criteria inventory classification using a new method of evaluation based on distance from average solution (EDAS). *Informatica*, 2015; 26(3):435–451.

[15] Keshavarz-Ghorabaee, M., et al. Multi-criteria inventory classification using a new method of evaluation based on distance from average solution (EDAS), Informatica, 2015; 26(3):435–451.

[16] Levinskytė, A., Banionis, K., Geležiūnas, V. The influence of thermal bridges for buildings energy consumption of "A" energy efficiency class. *Journal of Sustainable Architecture and Civil Engineering*, 2016; 2(15):44–58.

[17] Aviža, D., Turskis, Z., Volvačiovas, R. Correlation analysis of thermo-insulation layer thickness and its payback period of the typical pitched roof detail. *Procedia Engineering. 11th International Conference on Modern Building Materials, Structures and Techniques (MBMST)*. Amsterdam: Elsevier Science Ltd., 2013; 57. ISSN: 1877-7058.

[18] Kaupienė, J., Aviža, D., Baltušnikienė, R., Kasperiūnienė, Z. Efficiency of the wall-window linear thermal bridge in the net-zero energy building. *12th International Conference on Intelligent Technologies in Logistics and Mechatronics Systems, ITELMS'2018*, 26–27 April, 2018, Panevėžys, Lithuania / Koczy, L., Žostautienė, D., Strikulienė, O., Zacharovienė, E. Bologna: Editografica. ISSN 2345-0088. eISSN 2345-0096. 2018; pp. 89–97.

[19] Lu, Y., Wang, Sh., Zhao, Y., Yan, Ch. Renewable energy system optimization of low / zero energy buildings using single-objective and multi-objective optimization methods. Energy and Buildings, 2015; 89:61–75.

[20] Marszal, A. J., Heiselberg, P., Bourrelle, J. S., Musall, E., Voss, K., Sartori, I., Napolitano, A. Zero energy building – a review of definitions and calculation methodologies. *Energy and Buildings*, 2011; 43:971–979.

[21] Gatt, D., Caruana, C., Yousif, C. Building energy renovation and smart Iintegration of renewables in a social housing block toward nearly-zero energy status [Internet]. *Frontiers in Energy Research*, 2020; 8:560892. Available from: doi:10.3389/fenrg.2020.560892 (accessed on 01-09-2020).

[22] Rook wool [Internet]. *Catalog*. Available from: www.paroc.lt (accessed on 01-09-2020).

Forum on Innovative Technologies and Management for Sustainability – ITMS'2020

ktu 1922 panevėžys faculty of technologies and business

International Scientific Conference Intelligent Technologies in Logistics and Mechatronics Systems – ITELMS'2020, 1st October, 2020, Panevėžys, Lithuania

Classification of Surface Defects of Rolled Metal Using Deep Neural Network ResNet50

Ihor Konovalenko[a], Volodymyr Hutsaylyuk[b]*, Pavlo Maruschak[a]

[a]*Ternopil National Ivan Puluj Technical University, Department of Industrial Automation, 56 Ruska St, Ternopil 46001, Ukraine*
[b]*Military University of Technology, Faculty of Mechanical Engineering, 2 Gen. S. Kaliskiego St, Warsaw 00-908, Poland*

Abstract

An automated method for detecting and classifying several types of defects of rolled metal has been developed, which allows for conducting defectoscopy with the specified parameters of efficiency and speed of response. The method is based on the ResNet50 neural network classifier. A classifier has been trained to detect damage of three classes on flat surfaces with an overall accuracy of 95.8 % on the test data. ResNet50 has been shown to provide excellent recognition, high performance and precision, which makes it an effective tool for detecting defects on metal surfaces.

Peer-review under responsibility of the Kaunas University of Technology, Panevėžys Faculty of Technologies and Business.

Keywords: defect inspection; image processing; feature extraction; classification methodsdefects of rolled metal.

Introduction

In the steel industry, namely, in the process of sheet metal production, the task of controlling sheet metal to detect technological defects arises. Such control of steel products is one of the determining factors in evaluating their quality, reliability and suitability for use [1]. The existing automated systems for optical-digital control of the surface of sheet billets provide for the timely detection of defects. They are important because they allow diagnostics to be performed in parallel with the main technological operations [1–2]. This allows for the timely detection of external signs of billet damage and the adjustment of process parameters. Particularly convincing results in the field of image analysis are demonstrated by deep neural networks which allow for the recognition of objects on the images.

Defectometric surface control systems are manufactured by ISRA VISION PARSYTEC AG (Germany);

* Corresponding author.
E-mail address: volodymyr.hutsaylyuk@wat.edu.pl

DX20F0005

COGNEX CORPORATION (USA), Surface Inspection (UK), and others. The use of deep neural networks for the recognition of multiple surface defects of metal structures allows obtaining large amounts of information on the condition of the diagnosed object based on precision image recognition [3–5]. For various subject areas and tasks, there are arsenals of specialized problem solving methods – for example, ARMA, ARIMA, Box-Jenkins, and others.

For storage, processing, presentation and analysis of such data, an information-analytical system (IAS) for the morphology analysis of dimples of ductile tearing by means of neural networks was developed at the Ternopil Ivan Puluj National Technical University with the participation of the author [6]. The experience gained over the period of the system operation indicates the need to review some approaches to the analysis of optical-digital control data. Practice has shown that in order to estimate the condition of rolled metal based on the data of opto-digital control, it is necessary to develop new parameters for estimating the defective condition of the analyzed surfaces and finding indicators typical of such condition. One of the advantages of neural networks is that many of its elements can function in parallel, thereby significantly increasing the efficiency of work, especially in image processing. At the same time, neural networks are more robust than other statistical methods of image recognition, if the input image is noisy. Neural networks can serve as a basis for the development of automated diagnostics systems for metal structures. Therefore, the development and study of diagnostic systems based on image analysis using deep neural networks is an urgent task.

The purpose of this research was to develop a method for detecting and classifying defects by means of neural networks.

Defects and their classification. Defects of rolled metal are known to be standardized. In accordance with GOST 21014-88 [7], 64 types of black rolled metal defects are described and illustrated. Automated analysis of the defects of rolled metal allowed determining the morphological features and establishing the technological causes of the defects [7, 8] (Table 1).

Table 1. Defectoscopic signs of defects

Class	Defect features	Defect type	Reasons for occurrence
1	Single and multiple defects in the form of rounded spots	Shells	Shells can occur as a result of inclusions directly under the surface. Such inclusions are considerable stretched by subsequent deformation and tear open or are rolled over
2	Multiple linear and elongated defects of any orientation throughout the surface of rolled metal	Grooves / Scratches	These forms of damage occur as a consequence of relative movements between the rolled product and parts of the installation
3	Defects, which are peeling of a metal of a tongue-like shape. With separation of exfoliation, a deep depression with an uneven bottom	Rolling skin	It is the result of wear of the rollers

It should be noted that defects, even within a single class, might differ in shape, appearance, structure, which complicates the task of their unambiguous classification. Sometimes the defects of the same class are similar to those of another class, which requires accurate grading of the training sample. Neural networks allow using sequential training procedures, high generalization capabilities of diagnostic parameters, and the ability to analyze complex nonlinear links.

Train dataset. Based on the photographs of flat steel surfaces, a training sample containing 87 704 images of three damage types and images of undamaged surfaces was prepared (Fig. 1). Images are 256 × 256 pixels in size. Some of the images are from the Kaggle competition "Severstal: Steel Defect Detection" [9]. In general, the training sample contains 1 820 images of class 1 damage, 14 576 images of class 2 damage, and 2 327 images of class 3 damage. Some images show multiple class damage. There are 63 images with class 1 and 2 damage simultaneously, and 228 images with class 2 and 3 damage in the training database. The training dataset also contains 69 272 images of undamaged surfaces, which form 79 % of its volume. Thus, the training dataset is significantly unbalanced in terms of the distribution of images of different classes, which represents the distribution of classes in reality.

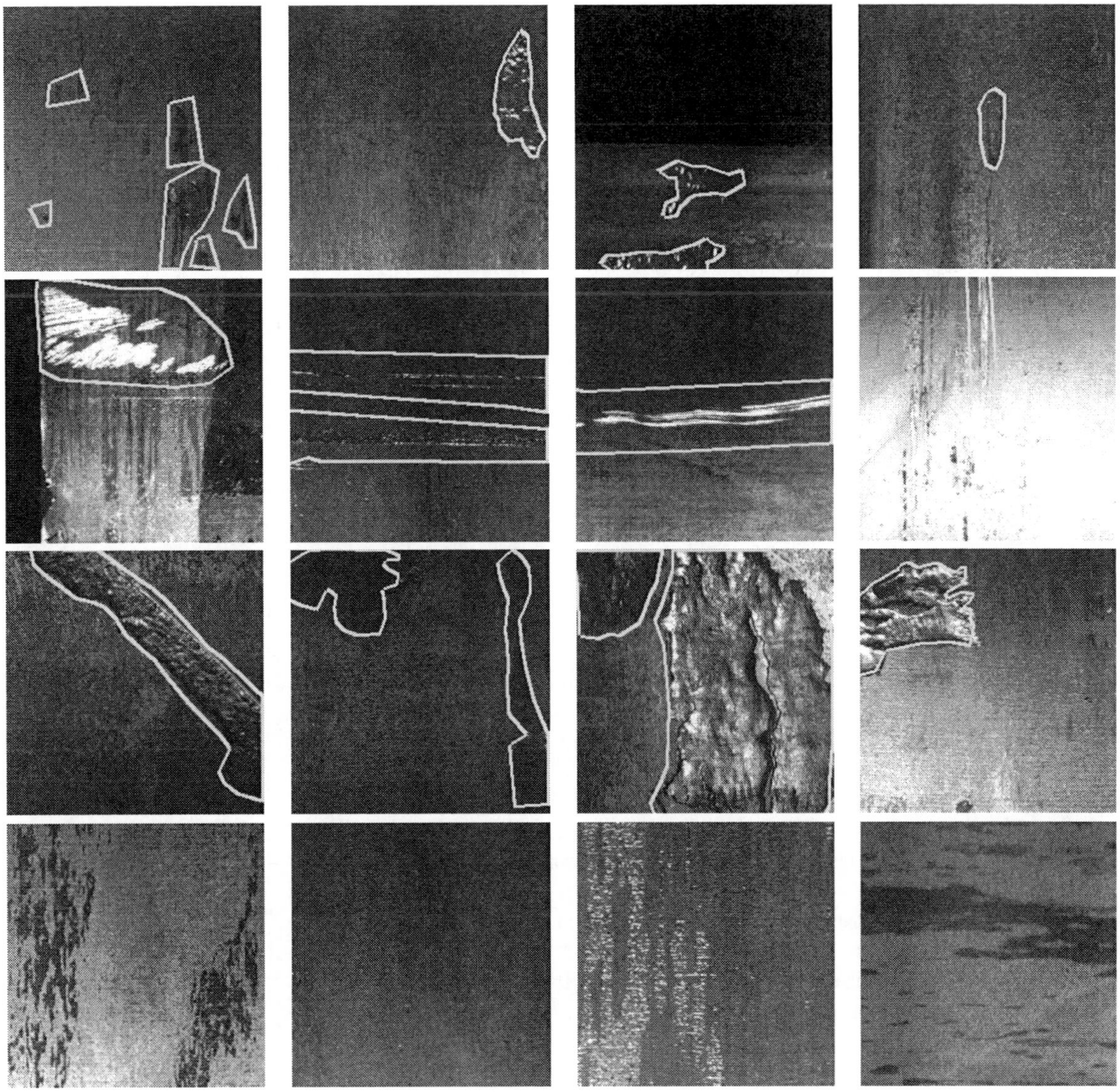
a
b
c
d

Fig. 1. Images with class 1 (a); class 2 (b); class 3 (c) damage (Table 1); and images of intact surfaces (d)

The surfaces of the steel strips under study has a different texture and shade, and the photos were obtained in different light conditions. Thus, the training sample is characterized by an appreciable variety not only of defects, but also of surfaces and the degrees of their illumination. Augmentation techniques were used to increase the diversity of the training dataset. To this end, the data generator, which supplied data to train the model, flipped the images with certain probability relative to the horizontal and vertical axes and rotated them at an angle multiple of 90°.

1. Classifier architecture

As backbone for classifier used residual neural network ResNet50 [10] presented in 2015 by Microsoft Research command. The ResNet50 demonstrated excellent generalization performance with fewer error rates on image recognition tasks and is therefore a useful tool for classification.

Architecture of ResNet50 allows to train extremely deep neural networks. Classifier has 50 stacked layers and over 23 million trainable parameters (FLOPs). The authors of [11] argue that stacking layers shouldn't degrade the network performance, because we could simply stack shortcut connections (layer that doesn't do anything) upon the current network, and the resulting architecture would perform the same.

The architecture of ResNet50 is shown in Figure 2. Model consists of 4 stages each with a residual block. Each residual block has 3 layers that make 1 × 1 and 3 × 3 convolutions. To reduce the problem of vanishing gradient in deeper layers ResNet network uses shortcut connections. Shortcut connections transmit the input directly to the end of residual block.

ResNet50 model makes the initial convolution and max-pooling using 7 × 7 and 3 × 3 kernel sizes respectively with stride 2 in both cases. Then stage 1 of the network starts and it has 3 residual blocks containing 3 layers each. The size of kernels used to perform the convolution operation in all 3 layers of the block of stage 1 are 64, 64 and 256 respectively. As we go from one stage to another, the channel width is doubled and the size of the input is reduced to half. The curved lines refer to the shortcut connection. The dashed curved lines represents convolution operation in the residual block that is performed with stride 2, hence, the size of input will be reduced to half but the channel width will be doubled.

As showed at Figure 2, ResNet50 uses bottleneck design in it blocks. For each block, three layers are stacked one over the other. The three layers are 1 × 1, 3 × 3, 1 × 1 convolutions. The 1 × 1 convolution layers reduce and then restore the dimensions. The 3 × 3 layer is left as a bottleneck with smaller input / output dimensions.

At the end the network has an average pooling layer followed by a fully connected layer having 3 neurons (by three classes of defects investigated).

The classifier is designed by using *Keras* and *TensorFlow* libraries.

2. Training

For training the classifier, we used transfer learning technique. As base took ResNet50 wages trained on a 1.4 million labeled images of 1000 classes from the ImageNet database. Using a pretrained model is a highly effective approach that allows use much less data for training.
All images were divided into test dataset (20 % of the total amount) and training dataset (the remaining 80 %). When training the model, 20 % of the training dataset was used for validation. Damage of all classes is presented in each group in proportion to its share in the total amount.

Two types of classifiers were trained: multilabel and multiclass. The multilabel classifier assumes that damage of several classes can be present in one image, while the multiclass classifier implies that each

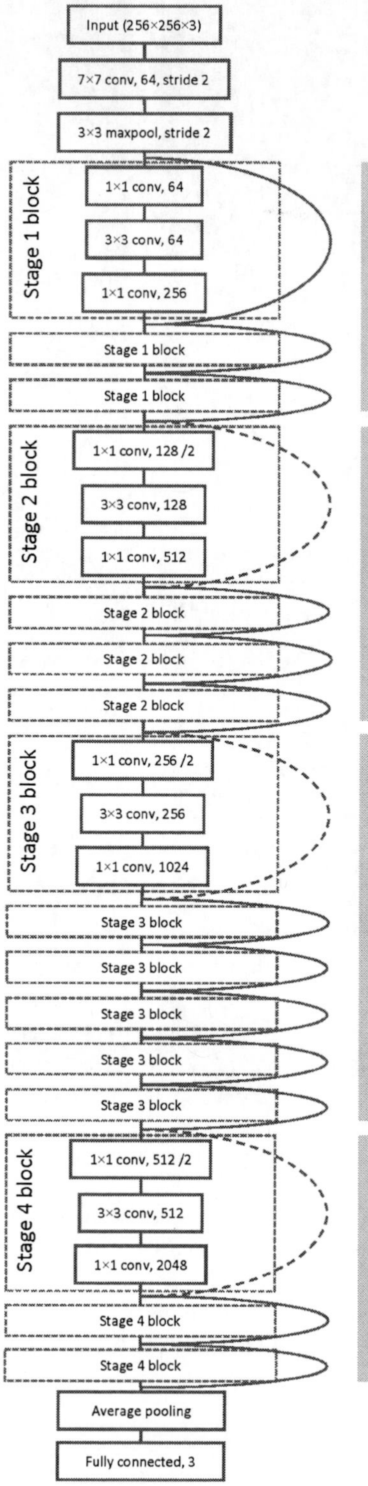

Fig. 2. ResNet50 classifier architecture

image represents only damage of one class. In our case, only 0.3 % of images contain damage of several classes, therefore, their total contribution to the error (if damage of only one class is found) is negligible. The output of the multilabel classifier contained 3 neurons, each of which signaled the presence (or absence) of damage of the relevant class. There were 4 neurons at the output of the multiclass classifier (one per each damage class and one for the images of intact surfaces).

When training multilabel classifiers, we used the binary crossentropy and binary focal loss functions. For multiclass variants, categorical crossentropy and categorical focal loss functions were used. At the end of each epochs, the following metrics were stored: false positives, false negativs, true positives, true negatives, accuracy, precision, recall and AUC (area under receiver operating characteristic curve).

The training process was studied with Adam, RMSprop and SGD optimizers. The learning rate was 0.001 at the beginning and decreased by 25 % after every 8 epochs, if the loss function did not improve its value. The batch size at training was 16 and 20 images. Training was performed until the loss function improved its value over 10 epochs. As a result of training, the model was selected, for which the loss function calculated based on the validation data was the lowest.

Training was performed using GPU NVIDIA GeForce GTX 1060 with 6 GiB inboard memory.

It was found that the best result achieved by applying the focal loss functions that proved to be suitable for unbalanced data [10]. Focal loss applies a modulating term to the cross-entropy loss in order to focus training on hard negative examples. It down-weights the well-classified examples and puts more training emphasis on that data that is hard to classify. In a practical setting where we have a data imbalance, majority class will quickly become well-classified since we have much more data for it. Thus, in order to ensure that we also achieve high accuracy on our minority class, we can use the focal loss to give those minority class examples more relative weight during training.

2.1. Multilabel classifier

When training a multilabel classifier, such hyperparameters as learning rate, batch size, and steps per epoch were changed. To evaluate the quality of a multilabel classifier, the binary accuracy calculated. Ground truth vector, which is fed to the output of the multilabel classifier during training, has the form $y^{gt} = \left[y_0^{gt}, y_1^{gt}, \dots, y_i^{gt}, \dots, y_{n_{cl}-1}^{gt}\right]$, where n_{cl} is the number of classes (in our case, $n_{cl} = 3$), y_i^{gt} is the output vector element, which equals 1, if the class with index i is present in the input image, or is 0, if not so.

The output vector with anticipation has the form $y^{pr} = \left[y_0^{pr}, y_1^{pr}, \dots, y_i^{pr}, \dots, y_{n_{cl}-1}^{pr}\right]$, where $y_i^{pr} \in [0; 1]$. If the element of this vector is $y_i^{pr} \geq t_b$ (t_b is a certain limit value), then the class with index i is present in the image, and vice versa. We define a step function that describes the following condition:

$$T_{bin}\left(y_i^{pr}\right) = \begin{cases} 0, & y_i^{pr} < t_b \\ 1, & y_i^{pr} \geq t_b \end{cases}; \tag{1}$$

We also define the equality function of two arguments, which returns 1 if they are the same, and 0 in the opposite case:

$$Eq(x_1, x_2) = \begin{cases} 0, & x_1 \neq x_2 \\ 1, & x_1 = x_2 \end{cases}; \tag{2}$$

Then, the binary accuracy for a single anticipation is:

$$a_{bin}^1 = \frac{1}{n_{cl}} \sum_i Eq\left(T_{bin}\left(y_i^{pr}\right), y_i^{gt}\right). \tag{3}$$

The binary accuracy of anticipation of the batch of input images with the size n_b has the form:

$$a_{bin} = \frac{1}{n_b} \sum_j a_{bin\ j}^1. \tag{4}$$

The best result was shown by the model trained with the SGD optimizer with the moment 0.9 at the learning rate 0.001, batch size 20, and count of steps per epoch 3000. Figures 3 (a, b) show how the binary accuracy of the classifier and the loss function changed during training

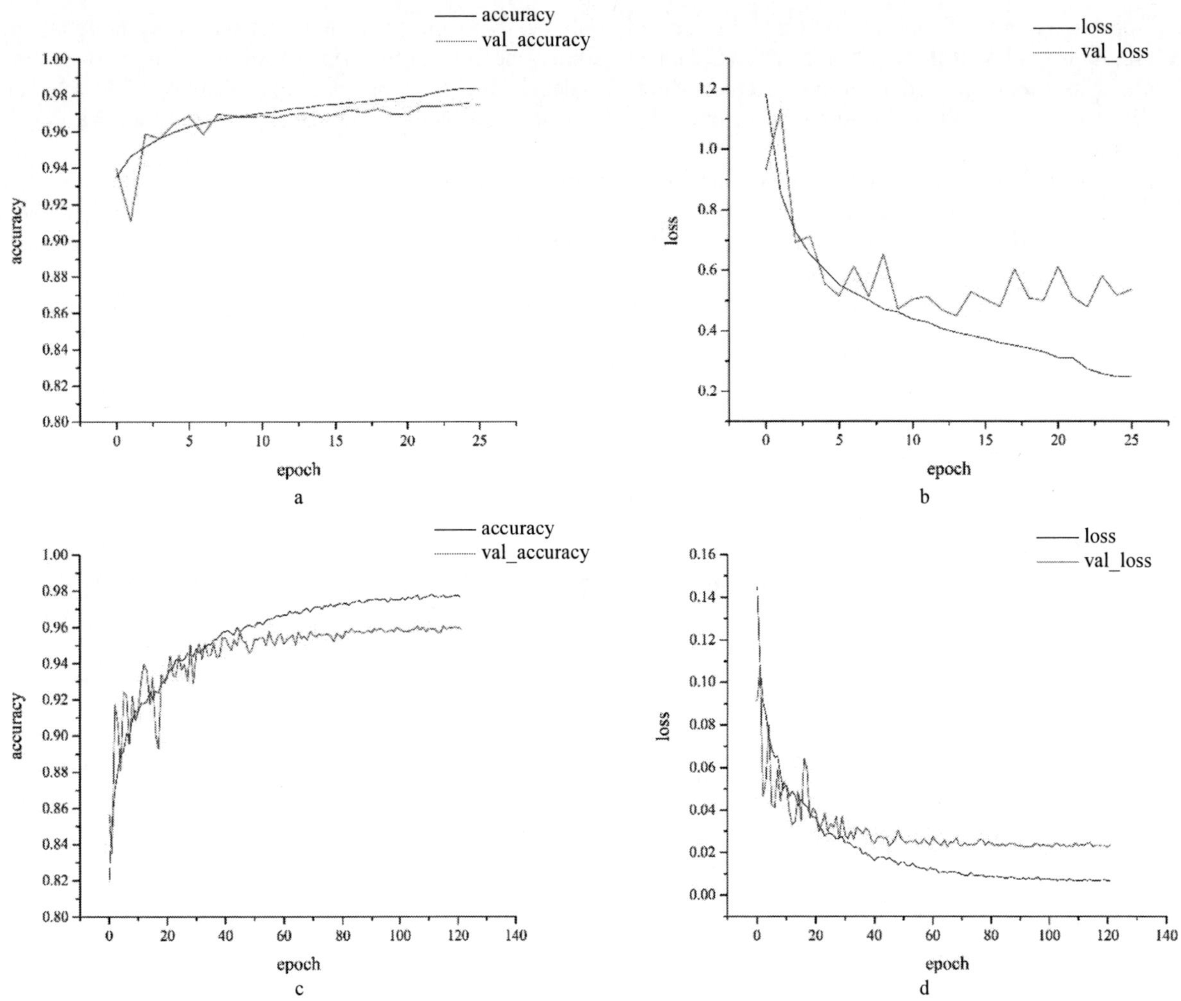

Fig. 3. Training curves of the multilabel classifier (a, b) and multiclass classifier (c, d). Variation of the accuracy metric (a, c) and loss function (b, d)

2.2. Multiclass classifier

When training a multiclass classifier, such hyperparameters as learning rate, batch size, steps per epoch, and class weights were changed. To estimate the accuracy of the model, categorical accuracy was calculated in addition to binary accuracy. The true value at the output of a multiclass classifier is the same as that of the multilabel classifier: $y^{gt} = \left[y_0^{gt}, y_1^{gt}, \ldots, y_i^{gt}, \ldots, y_{n_{cl}-1}^{gt}\right]$, however, for each value of this vector, only one element y_i^{gt} may equal 1, which indicates the presence of class *i* object in the image. Then, the categorical accuracy is:

$$a_{cat} = \frac{1}{n_b} \sum_j Eq\left(argmax(y^{pr}), argmax(y^{gt})\right). \quad (5)$$

The best result was achieved when using the SGD optimizer with the moment 0.9 at the learning rate 0.001, batch size 16, and count of steps per epoch 2000. In this case, we used the class weights calculated on the number of images of a particular class in the training dataset. This model trained much longer than others, but, at the same time, it has shown the best indicators of accuracy, recall and precision. The learning history of the abovementioned multiclass model is shown in Figure 3 (c, d). The quality metrics of the best developed models based on the ResNet50 neural network are summarized in Table 2.

Table 2. Quality metrics for classifiers

Model	Class	Binary accuracy	Categorical accuracy	Recall	Precision
Multilabel 1	1	0.9839		0.3018	0.8430
	2	0.9349		0.7403	0.8561
	3	0.9884		0.6981	0.8859
	Not damaged	0.9251		0.9658	0.9406
	Average	0.9581		0.6765	0.8814
Multilabel 2	1	0.9830		0.3509	0.7143
	2	0.9321		0.7346	0.8423
	3	0.9866		0.6169	0.8906
	Not damaged	0.9192		0.9392	0.9594
	Average	0.9552		0.6604	0.8517
Multiclass 1	1		0.9813	0.0723	0.9583
	2		0.9189	0.7656	0.7580
	3		0.9836	0.3604	0.8526
	Not damaged		0.9090	0.9614	0.9258
	Average		0.9482	0.5399	0.8737
Multiclass 2	1		0.9852	0.2759	0.9565
	2		0.9344	0.7208	0.8669
	3		0.9888	0.6873	0.8261
	Not damaged		0.9199	0.9788	0.9239
	Average		0.9571	0.6657	0.8934

Conclusions

Multilabel classifier 1 showed the best results for all classes, and its average accuracy is 0.9581 for all types of surfaces. Of the multiclass models, the best result was shown by the model, for which weights that adjust the distribution of samples by class were set during training.

It was found that damage of class 1, which most closely resembles surface formations typical of intact specimens, is the hardest to recognize. The multilabel classifier also allows adjusting the output for each class individually by selecting a threshold, at which the optimal quality of recognition is achieved.

The developed automated method for detecting and classifying three types of defects of rolled metal provides for the possibility of conducting defectoscopy with the specified parameters of efficiency and speed of response.

References

[1] Neogi, N., Mohanta, D.-K., Dutta, P.-K. Review of vision-based steel surface inspection systems [Internet]. *EURASIP Journal on Image and Video Processing*, 2014; 50:1–19. Available from: doi:10.1186/1687-5281-2014-50 (accessed on 18-06-2020).

[2] Sun, X., Gu, J., Tang, S., Li, J. Research progress of visual inspection technology of steel products – a review [Internet]. *Applied Sciences*, 2018; 8(11)-2195:1–25. Available from: doi:10.3390/app8112195 (accessed on 18-06-2020).

[3] Liu, Y., Xu, K., Xu, J. An improved MB-LBP defect recognition approach for the surface of steel plates [Internet]. *Applied Sciences*, 2019; 9(20):4222. Available from: doi:10.3390/app9204222 (accessed on 18-06-2020).

[4] Yi, L., Li, G., Jiang, M. An end-to-end steel strip surface defects recognition system based on convolutional neural networks. *Steel Research International*, 2017; 88(2):60–68.

[5] Xian, T., Dapeng, Z., Wenzhi, M., Xilong, L., De, X. Automatic metallic surface defect detection and recognition with convolutional neural networks [Internet]. *Applied Sciences*, 2018; 8(9):1575. Available from: doi:10.3390/app8091575 (accessed on 18-06-2020).

[6] Konovalenko, I., Maruschak, P., Brezinová, J., Brezina, J. Morphological characteristics of dimples of ductile fracture of VT23M titanium alloy and identification of dimples on fractograms of different scale [Internet]. *Materials*, 2019; 12:2051. Available from: doi:10.3390/ma12132051 (accessed on 25-06-2020).

[7] GOST 21014-88 *Rolled Products of Ferrous Metals. Surface Defects. Terms and Definitions.* Moscow, Izd. Stand., 1989; 61 p.

[8] Becker, D., Bierwirth, J., Brachthäuser, N., Döpper, R., Thülig, T. *Zero-Defect-Dtrategy in the Cold Rolling Industry. Possibilities and Limitations of Defect Avoidance and Defect Detection in the Production of Cold-Rolled Steel Strip.* Fachvereinigung Kaltwalzwerke e.V., CIELFFA, Düsseldorf, 2019; 16 p.

[9] Kaggle. *Severstal: Steel Defect Detection. Can you Detect and Classify Defects in Steel?* [Internet]. 2019. Available from: https://www.kaggle.com/c/severstal-steel-defect-detection (accessed on 25-06-2020).

[10] He, K., Zhang, X., Ren, S., Sun, J. Deep residual learning for image recognition [Internet]. *Proceedings of the IEEE Conference on Computer Vision and Pattern Recognition* (27–30 June, 2016), 2016; pp. 770–778. Available from: doi:10.1109/CVPR.2016.90 (accessed on 18-06-2020).

[11] Lin, T.-Y., Goyal, P., Girshick, R., He, K., Dollar, P. Focal loss for dense object detection [Internet]. *Proceedings of the IEEE International Conference on Computer Vision* (22–29 October, 2017), 2017; 2999–3007. Available from: doi:10.1109/ICCV.2017.324 (accessed on 25-06-2020).

Forum on Innovative Technologies and Management for Sustainability – ITMS'2020

panevėžys faculty of technologies and business
1922

International Scientific Conference Intelligent Technologies in Logistics and Mechatronics Systems – ITELMS'2020, 1st October, 2020, Panevėžys, Lithuania

Influence of Moisture on Thermal Conductivity of Mineral Wool

Jovita Kaupienė[a, b]*, Aurimas Česnulevičius[a, b]

[a]*Kaunas University of Technology, Panevėžys Faculty of Technologies and Business, 33 Nemuno St, Panevėžys LT-37164, Lithuania*
[b]*Panevėžys University of Applied Sciences, Faculty of Technological Sciences, 23 Laisvės Sq., Panevėžys LT-35200, Lithuania*

Abstract

The main indicator of thermal conductivity of building thermal insulation materials is the coefficient of thermal conductivity, the values of which are influenced by the properties of the physical material and environmental factors such as humidity and temperature. The relationship between the moisture content in stone wool and the thermal conductivity coefficient is investigated in this work using the experimental method. The changes of thermal conductivity coefficients with increasing soaking of stone wool in water are quantified as well in this work. There was estimated that with increasing water content in the investigated stone wool, the thermal conductivity also increases rapidly. Based on the experimental data, it was found that there is a functional relationship between the thermal conductivity coefficient λ and the long-term impregnation (relative humidity W_p).

Peer-review under responsibility of the Kaunas University of Technology, Panevėžys Faculty of Technologies and Business.

Keywords: thermal conductivity; mineral wool; coefficient of thermal conductivity.

Introduction

The construction sector is one of the major consumers of energy worldwide. The greatest potential for reducing greenhouse gas emission has been found to be in the construction sector. Residential energy consumption accounts for about 40 % of the European Union's (EU) final energy balance and Green House Gas (GHG) emission.

The vast majority of buildings in Lithuania consist of old residential houses. The energy efficiency of most buildings is extremely low: expensive heating, damp and cool rooms, the outdated appearance of the building, the non-ergonomic operation of the premises. The most popular way to renovate buildings currently is to insulate the building and renovate the facade.

---

* Corresponding author.
E-mail address: jovita.kaupiene@ktu.lt

DX20F0006

One of the most important properties of building thermal insulation materials is good thermal insulation. The main indicator of the thermal conductivity of the material is the coefficient of thermal conductivity of the material, the values of which depend on the physical properties of the material and environmental factors such as humidity and temperature. Hagentoft states that moisture is one of the main factors that directly or indirectly influences many decisions during the construction process [1].

One of the most effective thermal insulation materials is stone wool, which is widely used for insulation of building external partitions due to its properties such as low thermal conductivity, good fire and sound insulation and abundant raw material base, wide range of product applications, etc. [2]. Mineral wool (stone, glass) is the most popular thermal insulation material, accounting for a large share (60–80 %) of the market for these materials in many countries [3].

A large part of the scientific research of mineral wool was carried out in the eighties and nineties of the 20th century [4]. Thermal insulation materials in the external partitions of buildings are constantly affected by different climatic factors such as humidity, precipitation, wind, changes in cold cycles and various loads. Mineral wool products are often improperly installed and maintained (especially in flat roof constructions). This increases heat loss and has a negative destructive effect on the partitions [5]. Kiunzel [6] analogously to Buska states that water has a negative effect on the structures of buildings from the mechanical point of view and influences the deterioration of the properties of thermal insulation materials in the structure of buildings.

1. Thermal conductivity of mineral wool

The studies of the effects of humidity and temperature on materials were carried out in the thirties of the 20th century [7]. Glaser in 1958 developed a method for calculating moisture-resistant walls based on diffusion. One of the first and most common macroscopic models to study the motion of moisture and temperature in materials is the Philip and De Vries model [7]. Philip and De Vries developed an equation that describes the motion of moisture and heat under temperature and humidity gradients. These models were designed to study humidity and its effects on the external partitions of buildings. This section describes the factors that influence the thermal conductivity values of mineral wool and also reviews several studies related to the topic.

Gailius and Vėjelis [8] state that the thermal conductivity of stone and glass wool depends on such factors as solid and gas phase volume ratio, shape of air inclusions (pores), their size and amount and location in the material structure. The more uniform the arrangement of pores and voids, the better the thermal insulation properties of the material, with a decrease in the density of light fibrous materials, and at the same time an increase in the air permeability of the material [8]. This structure of mineral wool leads to an increase in the values of the thermal conductivity coefficient, as the heat transfer by convection through the air gaps and voids in the structure increases. The value of the thermal conductivity coefficient also depends on the ratio of the dense carcass of the stone wool to the pores in it. The thermal conductivity of the air in the pores is quite low (λ = 0.024–0.03067 W/(m·K)), and the thermal conductivity of the carcass can be up to 300 times higher [8].

In materials such as mineral wool, the process of heat flow penetration is complex, and the speed and direction of flow is determined by closed air pores, their size, and fibre contact zones. The thermal conductivity of such materials is also affected by the arrangement of fibres in the product structure. Since air is the best insulator, the heat flow, trying to bypass it, changes direction, looking for the lowest resistance, and penetrates the solid part of the phase (fibre) [8]. Thus, the smaller the fibre thickness and pore size and the more fibre intersections, the better the heat insulation.

Bliūdžius, Samajauskas [9] investigated the thermal and humidity states of fibrous thermal insulation materials. The aim of the research is to determine the regularity of heat and moisture movement and accumulation in the research structure. Different combinations of fibrous thermal insulation and wind insulation materials were selected for the study, as well as different ventilation intensities. Moisture migration in a layer where condensing moisture accumulates or which is in contact with another wet material has been investigated experimentally. Thermal and humidity calculations and experimental studies of wall structures in a climatic chamber were used to investigate the freezing of moisture in stone wool when the surface temperature is negative. The study found that the outer layer of fibrous thermal insulation material up to 20 mm thick falls into the zone of negative temperatures, but moisture condenses only in the nearest 10 mm thick layer. As a result, during the heating season, the humidity of unventilated fibrous thermal insulation may increase by up to 15 % and the thermal conductivity R value may decrease by 10 % [9]. Based on the research data, it was found that in three-layer constructions without air gap, with a relatively low (compared to the inner layer) water vapor permeability finishing layer, it is necessary to estimate the increase in heat transfer coefficient due to moisture freezing in the stone wool layer. The article provides a recommendation to reduce the calculated thermal resistance of the structure or to apply an additional supplement to the thermal conductivity coefficient for the assessment of the effect of accumulated moisture at negative temperature. The assumption is also confirmed that when the air warms up, the moisture accumulated due to negative temperatures (on the inner surface of the outer wall layer) does not manage to be absorbed into the finishing layer material and therefore drains from its surface and moisturizes other structures.

Jiričkova, et al. [10] described the study of changes in thermal conductivity of different types of mineral wool products, as the water content of the materials varies. Changes in the thermal conductivity coefficients of mineral wool products and the dependence of these changes on the moisture content in materials were studied. The research was performed experimentally and using two analytical methods: using the Bruggeman homogenization equations and the Winer boundary method. The results obtained by all methods used in the study are compared. Mineral wool products with different water absorption (hydrophobic and water-absorbing) were studied.

2. Tested materials and test equipment and methodology

Paroc WAS 35t ventilated partition board was used for the tests. This plate was selected according to the technical capabilities of the equipment in the laboratory: the maximum thickness of the tested sample can be up to 6 cm to determine the thermal conductivity of the material.

To determine the impregnation of *Paroc WAS 35t* stone wool board was used methodology described in the standard LST EN 12087:2013 "Building thermal insulation products. Determination of long-term immersion of water after immersion" [12]. The thickness of the *Paroc WAS 35t* tested sample must be the same as that of the original product. The sample should be square surface of 200 ± 2 mm. 6 samples with a side length of 200 × 200 mm and the same thickness as the original product of 30 mm were cut from the test stone wool board. The samples are numbered from 1 to 6. The first three samples are numbered for complete immersion in water, the rest for partial immersion in water when only the lower surface of the sample is in contact with water.

Prior to testing, the samples were kept in an oven (at 25 °C) for about 24 hours to remove any moisture that had accumulated in the environment. According to the test procedure, the samples must be kept for a minimum of 6 hours at a temperature of 23 ± 5 °C. The study was performed at an average ambient temperature of 19 °C. When determining the long-term impregnation according to the standard LST EN 12087:2013, the ambient temperature should be 23 ± 5 °C, and in exceptional cases the test may be performed at a temperature of 23 ± 2 °C.

To determine the thermal conductivity of the material, the *PHYWE* manufacturer's *Cobra 4* stand is used to determine the thermal conductivity of solids.

According to the test procedure used to determine the absorption of the mineral wool (by partial or complete immersion), a balance is required to determine the mass of the samples to the nearest 0.1 g. *KERN EW2200-2NM* scales are used to determine the sample masses. A water vessel with a device (weight) to maintain a constant water level and a device to hold the sample in the desired position (device to hold the sample in the desired position shall not cover more than 15 % of the surface area of the sample), 23 ± 5 °C tap water, drainage equipment and ruler. The samples are dried using a low temperature oven *SNOL67/350* with digital thermostat *OMRON E5CK-T*. The oven maintains a temperature of +50 °C to +350 °C.

Method for the determination of long-term water absorption after immersion in water. The long-term soaking of rock wool after complete immersion in water is determined by measuring the change in mass of the samples when the fully immersed samples remain in the water for at least 28 days [12]. In accordance with the procedure for determining the long-term impregnation of mineral wool after immersion in water, given in the standard LST EN 12087:2013, it is recommended to express the long-term impregnation W_{lt} in parts per volume. In this study, it is more rational to express the long-term impregnation of mineral wool relative to mass rather than volume, as provided for in the standard LST EN 12087:2013. The impregnation of a material by mass W, % (referred to in some literature as the moisture, moisture or humidity of the material) is most commonly used to express the moisture content of building materials. The relative humidity of the material by weight is expressed as a percentage (%). It shows the proportion of moisture in the material by weight. The relative humidity of a material by mass is determined by dividing the mass of moisture in the material by the mass of the wet material and multiplying by 100 (formula (1)).

$$W_p = \frac{m_W - m}{m_W} \cdot 100; \tag{1}$$

where W_p is the relative humidity (mass humidity) of the sample, %; m_w is the mass of the soaked sample, kg; m is the dry mass of the sample, kg.

Method for determination of long-term water absorption after partial immersion in water. The procedure for the determination of long-term partial immersion in water is analogous to the method for the determination of long-term complete immersion in water. The main difference between these methods is that the soaking of rock wool after partial immersion in water is determined by measuring the change in mass of the sample with its underside in contact with water for 28 days, rather than fully immersion. In this study, changes in mass are recorded over shorter time periods.

Methodology for determining the thermal conductivity coefficient. The thermal conductivity is determined by comparison, so the methodology used in the study is based on Fourier law. The test to determine the thermal conductivity coefficient λ takes a long time, because when heating the sample on one side, a steady heat flow must develop through it, and only then the surface temperatures of the sample are measured and the heat flux is calculated.

The determination of the thermal conductivity coefficient by the said stand is based on the calculations of the heat flux (emitted by a 60 W incandescent lamp inside the thermal insulation house) passing through the stone wool board towards the outside. The heat flux from the heat source to the test plate and from the test plate to a lower temperature environment is expressed by equations (2) and (3):

$$Q = \alpha_1 \cdot (T_{f1} - T_{p1}) \cdot F; \tag{2}$$

$$Q = \alpha_2 \cdot (T_{p2} - T_{f2}) \cdot F. \tag{3}$$

where α_1 and α_2 are the heat transfer coefficients in the presence of natural air movement in the room (it was found experimentally that this coefficient is equal to 8.1 W/(K·m^2)); T_{f1} and T_{f2} are respectively the temperatures of a warmer insulating house interior and colder exterior, °K; T_{p1} and T_{p2} are respectively the temperatures of the warmer and colder surfaces of the examined plate, °K; F is the area of the examined plate, m^2.

The heat flux passing through the test stone wool board is described mathematically by the equation of thermal conductivity through the flat wall:

$$Q = \frac{\lambda}{\delta} \cdot (T_{p1} - T_{p2}) \cdot F; \tag{4}$$

where λ is the thermal conductivity coefficient of the test plate, W/(m·K); δ is test plate thickness, m; F is the area of the examined plate, m^2.

3. Analysis of stone wool impregnations

Information about samples (stone wool *Paroc WAS 35t*) impregnation is presented in Table 1.

Table 1. Mass and relative humidity of the samples material.

t, days	Sample No. 1		Sample No. 2		Sample No. 3		Sample No. 4		Sample No. 5		Sample No. 6	
	m, g	W_p, %	*m*, g	W_p, %	*m*, g	W_p, %	*m*, g	W_p, %	*m*, g	W_p, %	*m*, g	W_p, %
0	102.62	0	100.4	0	98.54	0	96.36	0	101.76	0	104.17	0
1	165.32	37.93	145.92	31.2	143.56	31.36	106.38	9.42	109.81	7.33	109.81	5.14
2	195.48	47.5	197.07	49.05	194.31	49.29	107.4	10.28	117.21	13.18	117.21	11.13
5	248.58	58.72	250.1	59.86	246.89	60.09	114.55	15.88	120.87	15.81	120.87	13.82
9	268.31	61.75	269.42	62.73	271.2	63.67	117.74	18.16	122.35	16.83	122.35	14.86
15	292.88	64.96	286.18	64.92	293.43	66.42	125.85	23.43	123.36	17.51	127.53	18.32
21	304.58	66.31	344.89	70.89	331.57	70.28	130.54	26.18	128.49	20.8	131.51	20.79
28	470.1	78.17	505.01	80.12	481.32	79.53	147.54	34.69	152.01	33.06	157.65	33.92

The results obtained during the study show that *Paroc WAS 35t* stone wool has a very high water absorption. It was found that the investigated rock wool can absorb water up to 5.03 times compared to the initial mass of the samples (sample No. 2 absorbed the most water).

The changes of samples impregnation over time are shown in Figure 1.

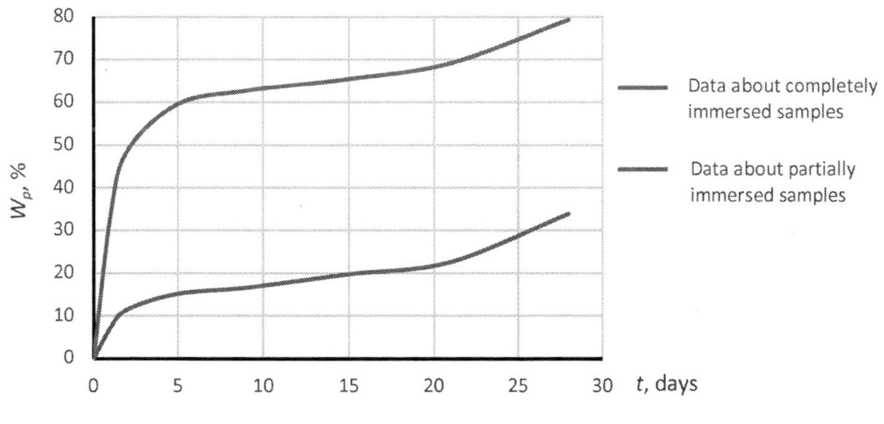

Fig. 1. Changes of samples impregnations over time

The results of the study show that the completely immersed samples No. 1, 2 and 3 achieves an average 2.34 times larger water absorption than partially immersed rock wool samples No. 4, 5 and 6. This difference is clearly seen in Figure 1.

Relative humidity W_p of completely immersed *Paroc WAS 35t* samples (No. 1, 2, 3) after a period of 28 days reaches an average value of 79.3 %. The fastest change in impregnation takes place during the first 5 days: samples completely immersed in water reach 76 % of the total impregnation. In addition, impregnation of sample No. 1 and 3 varies slightly from day 2 to day 6 and increases only 9 %.

Relative humidity W_p of partially immersed *Paroc WAS 35t* samples (No. 4, 5, 6) after a period of 28 days reaches 33.9 %. The fastest change in impregnation of the partially immersed samples occurs before the 4th day, when the samples reach 41 % of the relative humidity achieved during the test. The relative humidity of the samples No. 4–6 in the period from the 5th to the 20th day increases only 7 %.

Analysis of the relationship between impregnation and thermal conductivity. The obtained data show that with the increase of long-term impregnations of the samples (No. 1–6), their thermal conductivity coefficients also increase rapidly. The average change in thermal conductivity of all samples with increasing impregnation is shown in Figure 2.

After 28 days of immersion in water, the thermal conductivity coefficients of fully immersed samples are 1.98 times higher than those of partially immersed samples. It can be seen from Figure 2 that the average changes in thermal conductivity coefficients of fully and partially immersed water samples are very similar when the relative humidity W_p are relatively small and ranges from 0 % to 35 %. However, a comparison of the changes in thermal conductivity when relative humidity is greater than 35 % is impossible because relative humidity of the partially immersed samples does not reach 34 % throughout the test period. Thus, based on the results of the study, it can be assumed that in the case of small impregnations (W_p < 35 %) the thermal conductivity coefficients do not depend on the method of impregnation in water.

The thermal conductivity coefficient of *Paroc WAS 35t* rock wool samples (No. 1, 2, 3) completely immersed in water increased on average from 0.0333 W/(m·K) to 0.0977 W/(m·K), i.e. changed 0.0645 W/(m·K) or 2.94 times (average relative humidity reaches 79.27 %). The average change in thermal conductivity when moisture ammount in samples No. 1–3 increases by 1 % is 0.000813 W/(m·K) or 3.71 %.

The thermal conductivity coefficient of *Paroc WAS 35t* samples (No. 4, 5, 6) partially immersed in water increased on average from 0.0335 W/(m·K) to 0.0493 W/(m·K), i.e. changed 0.0158 W/(m·K) or 1.47 times. The average change in thermal conductivity when moisture ammount in samples No. 4 6 increases by 1 % is 0.00047 W/(m·K) or 4.35 %.

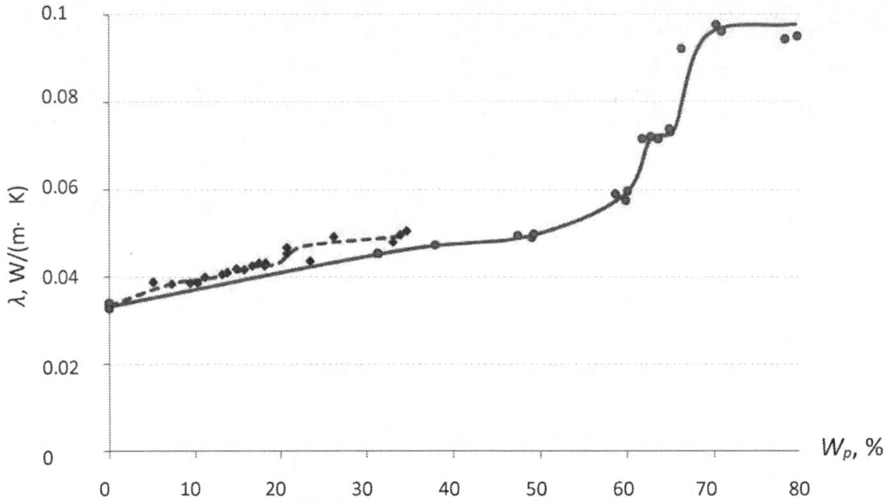

Fig. 2. Changes of the thermal conductivity coefficients λ with increasing impregnation (W_p) — the mean change of λ when the specimens are completely immersed, - - - – the mean change of λ when the specimens are partially immersed, ◆ – experimentally determined λ values for partially immersed samples, ● – experimentally determined λ values for fully immersed samples

The thermal conductivity coefficients of fully immersed tested stone wool samples (No. 1–3) reach values higher than the value of the thermal conductivity coefficient of water (0.60 W/m·K, 10 °C temperature [13]). This confirms the statements of researchers Jiričkova and et al. [10] that the values of thermal conductivity of wet mineral wool products are somewhere between the values of thermal conductivity of water and basalt (3.5 W/(m·K) [13]). During the soaking of stone wool the air in the pores is replaced by water and the thermal conductivity of the air is about 24 times lower (0.025 W/(m·K) [13]) than that of water (0.60 W/(m·K), 10 °C temp. [13]). As a result, the thermal conductivity coefficients of soaked stone wool increase as the amount of water in it increases.

Conclusions

1. The coefficient of the thermal conductivity of fully immersed tested stone wool samples increases on average from 0.0333 W/(m·K) to 0.0977 W/(m·K) over 28 days, i.e. changes 0.0645 W/(m·K) or 2.94 times and the relative humidity reaches 79.27 %.
2. The coefficient of the thermal conductivity of partially immersed tested stone wool samples increases on average from 0.0335 W/(m·K) up to 0.0493 W/(m·K), i.e. changes 0.0158 W/(m·K) or 1.47 times and the relative humidity reaches 33.89 %.
3. When the impregnation varies within $0 < W_p < 50$ %, the relationship between coefficient of thermal conductivity λ and the long-term impregnation (W_p) is described as linear.

References

[1] Hagentoft, C. E. et al. Assessment method of numerical prediction models for combined heat, air and moisture transfer in building components: benchmarks for one-dimensional cases. *Journal of Thermal Envelope and Building Science*, 2004; 27(4):327–352.

[2] Vėjelis, S., Vaitkus, S. *Termoizoliacinės medžiagos ir šiluminė varža* [Internet]. Available from: http://lt.lt.allconstructions.com/portal/categories/83/1/0/1/article/16041/termoizoliacines-medziagos-ir-silumine-varza-1 (accessed on 19-07-2020).

[3] Steponavičius, L. Ilgalaikėmis gniuždančiomis apkrovomis veikiamų akmens vatos plokščių deformatyvumo tyrimai [Internet]. *14-oji Lietuvos jaunųjų mokslininkų konferencija "Mokslas – Lietuvos ateitis"*, 2011; pp. 1–8. ISBN 978-9955-28-929-6. Available from: http://dspace.vgtu.lt/bitstream/1/770/1/4_Steponaitis-S1.pdf (accessed on 19-07-2020).

[4] Gurskis, V. *Statybinės medžiagos.* Mokomoji knyga. Kaunas: Ardiva, 2008; pp. 122–125.

[5] Buska, A. *Makrostruktūros ir kitų savybių įtaka mineralinės vatos gaminių gniuždymo rodikliams* [Internet]. Daktaro disertacija. Vilnius: Technikas, 2010; 206 p. ISBN 978-9955-28-590-8. Available from: http://vddb.library.lt/fedora/get/LT-eLABa-0001:E.02~2010~D_20100622_145715-68327/DS.005.0.01.ETD (accessed on 15-07-2020).

[6] Künzel, H. M. *Simultaneous Heat and Moisture Transport and Building Components. One-and-Two Dimensional Calculation Using Simple Parametres.* Fraunhofer Institute of Building Physics, Verlag Suttgart, 1995; 65 p. ISBN 3-8167-4103-7,

[7] Philip, J. R., De Vries, D. A. Moisture movement in porous materials under temperature gradients [Internet]. *Eos, Transactions American Geophysical Union,* 2014; 38(2):222–232. Available from: https://www.researchgate.net/publication/230606281_Moisture_movement_in_porous_materials_under_temperature_gradient (accessed on 20-07-2020).

[8] Gailius, A., Vėjelis, S. *Akustinės ir termoizoliacinės medžiagos* [Internet]. Mokomoji knyga, Vilnius, Technika, 2012; pp. 62–94. ISBN 978-609-457-395-8.

[9] Bliūdžius, R., Samajauskas, R. Peculiarities of determining thermal conductivity coefficient of low density fibrous materials. *Materials Science / Medžiagotyra.* Kaunas, Technologija, 2001; 7(4):280–284.

[10] Gimbutis, G. *Šiluminė technika.* Vilnius, Mokslas, 1993.

[11] LST EN 12087:2013 *Statybiniai termoizoliaciniai gaminiai. Ilgalaikės vandens sugerties panardinus jame nustatymas.*

[12] STR 2.01.03:2009 *Statybinių medžiagų ir gaminių šiluminių techninių dydžių projektinės vertės* [Internet]. Lietuvos Respublikos Aplinkos ministerija, Valstybės žinios Nr. 95-4047, 2009-07-29. Available from: http://www3.lrs.lt/pls/inter3/dokpaieska.showdoc_l?p_id=350732&p_ query=&p_tr2= (accessed on 19-07-2020).

Forum on Innovative Technologies and Management for Sustainability – ITMS'2020

panevėžys faculty of technologies and business

International Scientific Conference Intelligent Technologies in Logistics and Mechatronics Systems – ITELMS'2020, 1st October, 2020, Panevėžys, Lithuania

The Analysis of Building Waste Formation and the Possibilities of Its Reusage

Jovita Kaupienė[a, b], Agila Zalatorienė[b]*

[a]Kaunas University of Technology, Panevėžys Faculty of Technologies and Business, 33 Nemuno St, Panevėžys LT-37164, Lithuania
[b]Panevėžys University of Applied Sciences, Faculty of Technological Sciences, 23 Laisvės Sq., Panevėžys LT-35200, Lithuania

Abstract

In these times of modern technologies, fast production and consumption, it is impossible to life without waste. Therefore, people look for different ways how to recycle the waste, how to re–use it, but not to pile it in dumps or burn. In this work the structure of building waste, its amount and the possibilities of recycling this waste are analysis. Since 2014, the amount of building waste has been growing up. It is impossible to fix the real amount of building waste, because we lack the common system of registration. Concrete, bricks, tiles and ceramics from the part in the flow of buildings waste.

Peer-review under responsibility of the Kaunas University of Technology, Panevėžys Faculty of Technologies and Business.

Keywords: buildings waste; recycling; elimination; usage.

Introduction

The growing amount of energy and not renovated usage of materials create a threat to the environment. From this point of view, stable production (including building as well) is acknowledged the new industrial revolution non which needs a holistic attitude in order a living cycle of any product to be shut and it included different aspects of stability of the product in all its living stages. Most often people using recycling, recollection, re-usage and elimination for the last stage of using products (Fig. 1). These strategies are construction especially important in the field of (building) where building and demolition from 25–30 % of all possible waste, which is 500 mln. tons of waste in Europe.

European Parliament and Council adopted the directive 2008/98 EB [1] for arranging waste, also foresaw the means how to protect the environment and people's health. In order to realize the aims of this directive the member states of

* Corresponding author.
E-mail address: agila.zalatoriene@panko.lt

DX20F0008

the European Union should take means to achieve that up to 2020 of this at least 70 % of not dangerous building and demolition waste should be made ready to reuse or to recycle it. In many countries this index has been achieved, but in Lithuania, alas, not.

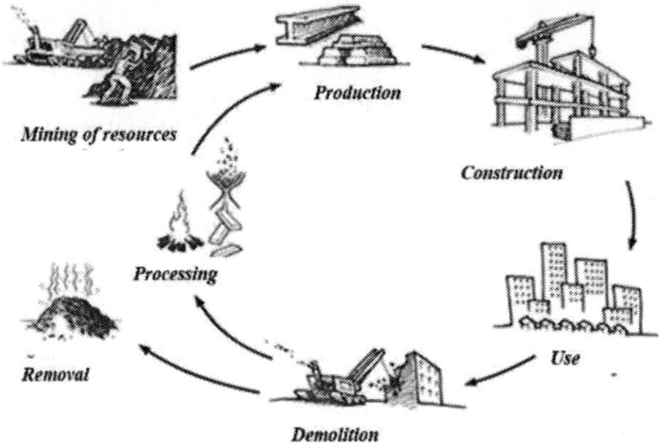

Fig. 1. Scheme of circular economy of building

1. Survey of buildings waste

The rules of arranging building waste [2] in Lithuania note that building waste is the waste which is made during the moment of building, reconstruction, repairs or demolition. "Waste Register" part 17 (Rules No. 217) name the codes of building and demolition waste (including the ground taken from contaminated places). These codes also include concrete, bricks, tiles and ceramics; wood, glass, plastic, bitumen mixture, coal rosin and tarred products; metals (including their alloys); ground (including the ground from contaminated places), stones and soaked out silt; isolation materials which contain asbestos; plaster, building materials of isolation. Some other building waste times building waste can be not dangerous, l. g. concrete or dangerous such as the mixture of concrete, bricks, tiles and ceramics or their separate fractions which contain dangerous materials.

The amount of building waste in different EU states, is not the same and it in difficult even to compare as the statistics of waste gathering is being led differently (different ways of collection, different waste is included into statistics) [3].

Figure 2 shows the formation of waste in EU members depending on the circulation of construction of industry (the amount of waste in tons the circulation industry in mln. euros). The greatest difference between the biggest and the smallest amounts of waste announced appear due to some facts, e.g. Finland stated huge amount of ground as building waste while Malta used the methodology which doubled the amounts of waste, Sweden showed the amount of building waste which was processed by the licensed plans.

In 16 October, 2016, European Commission presented the suggestion concerning the record of voluntary arrangement of building and demolition waste. The aim of this record is to perfect the establishment of waste, also logistics, recycling and governing the quality.

In order to produce second-rate materials of high quality it in important to sort and separate the sources, also logistics, recycling and governing the quality.

If we want to get high quality second-rate materials, we should pay attention to sorting and separating waste. In order to achieve good results in the process of sorting out building and demolition waste it is important for members to cooperate in building and recycling market, to have high recycling technologies as well as favorable markets to realize recycled waste.

Main unsuitable ways of arranging building waste:

- elimination of waste in dumps is the least suitable principle of elimination as the period of disintegration is long;
- pouring the waste into the least permitted places (water reservoirs, ditches, canals, forests) as the components of building waste can react to each other or with oxygen and from dangerous units;

- burning of waste as it is dangerous to health.

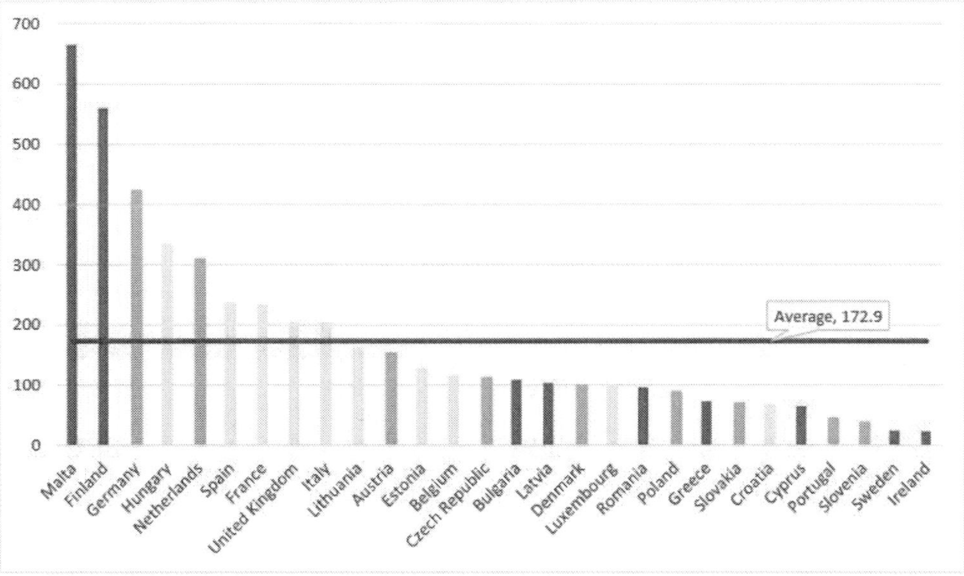

Fig. 2. The amounts of building waste in the states of EU (Units: the amount of waste in tons / circulation of building industry in mln. euros in 2016 [4])

While ordering or arranging waste one should keep to hierarchical arranging principles:

- responsible usage;
- avoiding waste;
- lessening the amount of waste;
- re-usage;
- recycling;
- usage to get energy;
- elimination in dumps.

According to the EU register dumping building waste in the least acceptable principle, however the accounts of Lithuanian Environment Protection. Agency show that at this time, the major part of the waste is dumped and only a small part is burned. Also, the hierarchy of waste in Lithuania is violated and therefore less waste reused repeated than recycled. Reusing and recycling waste in a building sector might lessen the need of primary materials as well as the cost of a building object. Also, if we integrate the repeated usage in the building sector, we might realize the principles of circular economy.

2. The analysis of the amount and composition of building waste in Lithuania

A lot of scientific research has been carried out while investigating the creation of new building materials or perfecting their industrial process, also trying to prolong the longevity of building materials, and lessen the building cost. It would be expedient to start our investigation from the analysis of the composition and amount or building materials if we want to lessen the amount building waste.

Since 2014 to 2017 the amount of building waste had been growing. The change of it can be seen in P.3 (according to the accounts of Environment Protection agency). We notice that the amount waste became bigger a year from 6 % to 16 % and already the amount of this waste reached 1 mln. tons in 2016. It should be noted that it has not been possible to fix the amount of mixed building waste as yet because the estimation of the market members of the whole amount of building and demolition waste has not been registered. The Biggest amount of building waste in Lithuania has been collected by the grounds of large size waste (the waste is in accepted from physical person and only in sorted fractions), by dump of safe waste (from physical and juridical person, some dumps accept mixed waste), by enterprises which have the legal right to sort, recycle safe inert building waste. One more flow of the waste is mixed communal waste. This waste contains at about 5 % to 23 % building waste in the general amount of mixed communal waste.

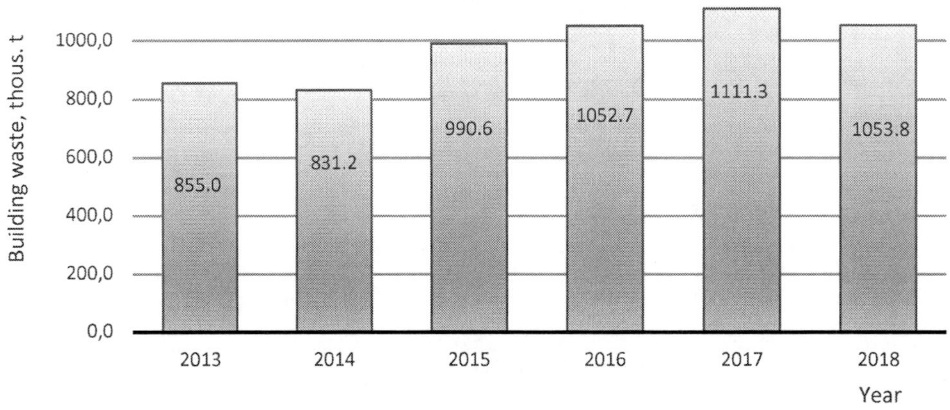

Fig. 3. The amounts of building waste in Lithuania in 2013–2018

This waste is not accounted or not accurately accounted for some reasons: first of all, because of unsuitable way of dumping (buried in forbidden places, emptied in ditches or forests, dumped in communal waste containers) also, because of the control being out of order or improper order. It is expected that the situation will change when GPAIS system is installed, while now it is stated that the amount building waste of in reality may as be half as bigger as the present amount.

In the flow of the building waste the part of the waste is occupied by the mixed building waste (Fig. 4). It forms from 26 % to 36 %, concrete forms about 25 %, metals about 13 % of the building waste.

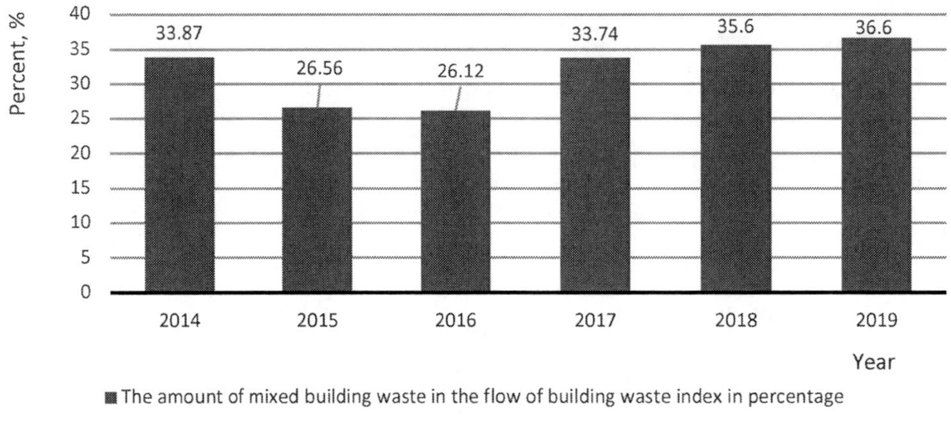

Fig. 4. The change of mixed building waste in the amount of building waste from 2014 to 2019 (the index of Region centres of managing waste)

As mixed building waste forms the biggest part in the flow of building waste, we can make conclusion that the waste on the building site is not sorted out, but most often is throw and mixed together with inert and insulating materials. This mixture of different sorts of waste makes a repeated usage of building waste especially difficult. It is absolutely impossible to get second-rate raw materials, from the mixed building waste flow or the index of quality makes their recycling repeated usage impossible.

The smallest part of building waste is made of dangerous waste which contain asbestos or plaster or with dangerous chemical materials. The amount with dangerous building waste gathered since 2008 (index of environment agency) can be seen in Figure 5. We can see that in the period of 2008 to 2014 the amount was very small, but since 2015 the number has grown several times. This has happened because since this time most dumps have stopped accepting mixed building waste and took only sorted in different fractions.

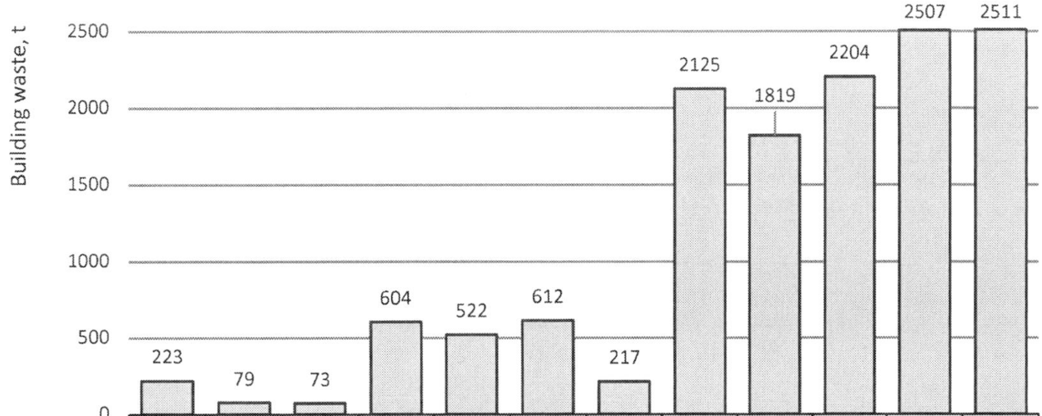

	2008	2009	2010	2011	2012	2013	2014	2015	2016	2017	2018	2019
Building waste, t	223	79	73	604	522	612	217	2125	1819	2204	2507	2511

Fig. 5. The amount of building and demolition waste, tons (the associations of the Regional centres of waste managing)

If we analyze the flow of mixed communal waste in Panevėžys district (Fig. 6), we can see mixed building waste at different season of a year. Building waste in mixed communal waste is the biggest in Panevėžys, Pasvalys and Kupiškis regions in winter, while in Rokiškis region and Panevėžys city – in autumn, whereas in Biržai region – in summer.

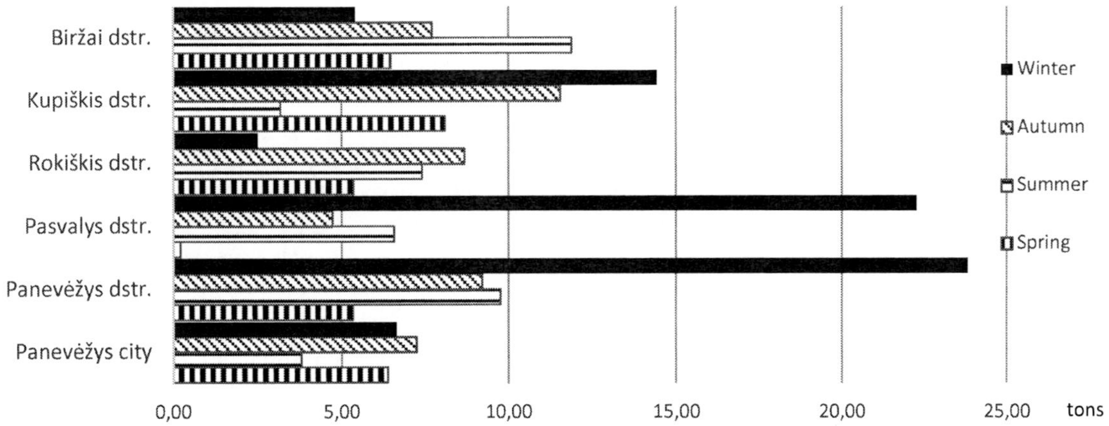

Fig. 6. The change in % of mixed building waste in the flow of the mixed communal waste in Panevėžys district in different season of year. Source of information: The data of the structure of safe waste being dumped or being accepted by plants of biological treatment in Panevėžys district

It is practically possible to process building waste (separated it in biological mechanical plants) repeatedly because it is polluted and wet so mostly it is dumped.

There are still quite a lot of houses in Lithuania, which roofs of asbestos slate. Asbestos is a cancerogenic material, though it is dangerous to people health only then it is being broken. In Figure 7 we can see the amount of asbestos containing waste which is collected.

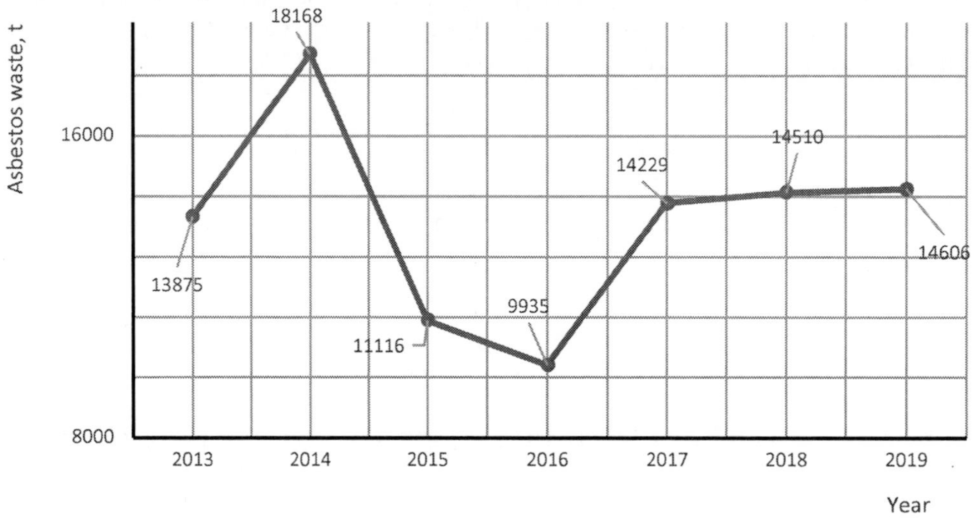

Fig. 7. The amount of asbestos containing waste which is collected from 2013 to 2019

In average, it is collected about 14000 tons a year. All this amount is removed in special sectors for gathering asbestos in dumps.

At this time, it is strictly forbidden to produce and market the products containing asbestos, but the process of excluding the products containing asbestos is rather slow.

In 2019 ministry of environment declared that it is possible that about one million tons of asbestos concrete products may have been used and 96 % of them are slate – covered roofs (containing asbestos).

3. Recycling of building waste and its repeatable usage

A lot of waste is formed while demolishing and reconstructing buildings. A big part of these materials can be recycled and used repeated, however the amount of recycled and reused materials differs a lot. According to the index of the municipalities of Lithuania one can notice the biggest part of building waste is recycled in Lithuania. Lithuania export only 16 % of processed materials from the flow of building waste. Mostly they are iron, capper, aluminum, bronze. The process in applied to the similar amount from the building waste, but only 10 % are reused. 4 % of the waste in dumped and even less, about 0.4 %, is burnt, that is used to produce energy. Such a situation happens to be because the burning Capacities in Lithuania are too small and the enterprises of such a type can accept a very small amount of waste to burn.

According to the state register of waste 121 enterprises in Lithuania have the right to recycle safe inert building waste. 63 enterprises can recycle concrete, 52 – bricks, 32 – tiles and ceramics. Building materials based on plaster and building materials based on asbestos are not recycled n Lithuania. Two enterprises recycled wood, 8 enterprises – glass, 2 enterprises – plastics, asphalt containing tar is not recycled, while tar and tar products are recycled, by 40 enterprises. Metals (including alloys), ground and pumped out silt, materials of isolation not are recycled in Lithuania.

Concrete, bricks, tiles and ceramics make the biggest pat part in the flow of building waste. They are also most easily recycled, that is ground, divided into fractions. The biggest part of different types of road metal is processed and it used in road building and renovating. Wood building waste perfectly suits for recycling and production of biomass energy. When plastics well separated it can be reused and recycled. Glass building waste can be used to repeated production of glass.

Figure 8 shows that concrete and mixed building waste make the biggest processed part in the flow of building waste. If we compare the years 2016, 2014, 2018, 2019, we can see that the general amount of processed building waste has not changed (including concrete, bricks, tiles, bitumen mixtures and road metal), only the amount of processed building waste in 2018 and 2019 has grown up three times compared to 2016 and 2017.

Making concrete waste (CW) small the produced material can be used instead of road metal when we need the concrete of small firmness (20 % weaker than it was before). Concrete waste is not classified, but put to the waste at once. Therefore when we produce concrete for fillings we can add not more than 15 % of this processed waste as ouch is the requirement in Lithuanian standards. Complicated regulations are one of the reasons why the amounts of processed building waste do not grow.

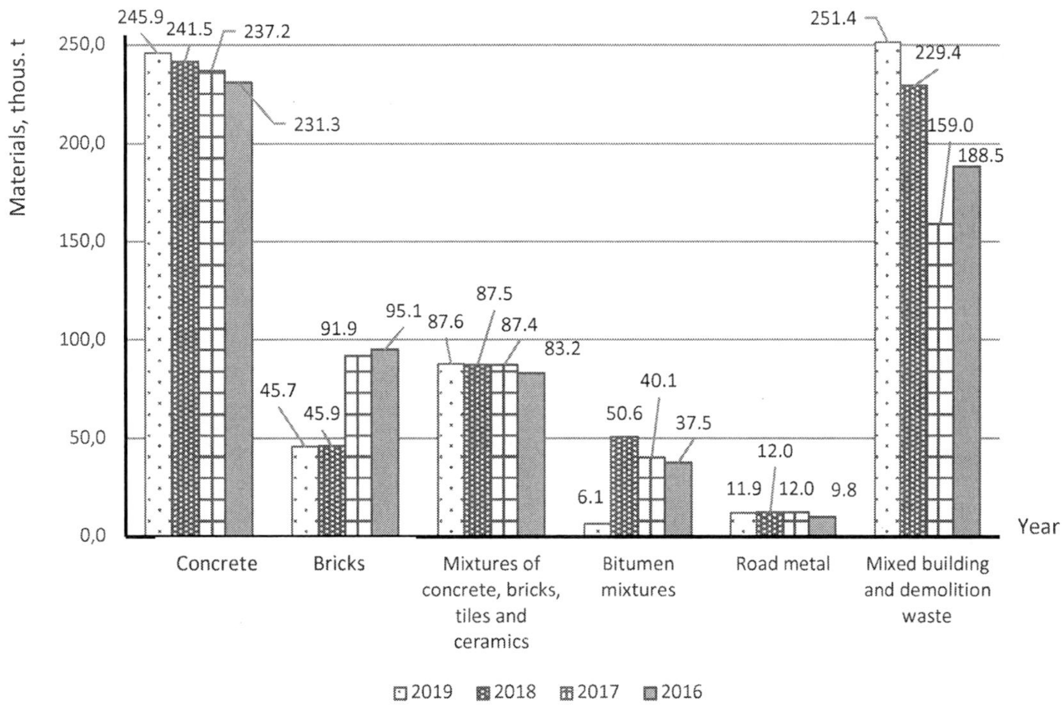

Fig. 8. The materials which are most often processed in the flow of building waste

Processed concrete can be applied to building or reconstructing roads. It is possible to mix the processed product with burnt wood ashes which can be used as building material or its substitute in civil engineering buildings.

Conclusions

The EU directive of 2008/98 promotes the aim that till 2020 70 % of waste, at least should be recycled. If we want to reach those goal, we should take use of the experience of those states of EU where building waste is managed best. For this purpose these is arranged the State program of waste prevention, the state plan of arranging waste.

The present waste politics of EU is based on the principle of hierarchy of waste arranging ways and it is violated in Lithuania, that's why the building sector, can't lessen the need of primary materials and the cost of building objects. If we could integrate repeated usage of waste in the building sector, we could manage to ensure the implantation of circular economy principles and lessen the amount of building waste in dumps.

The amount of building waste grows from 6 % to 16 % in 2014–2017 and in 2016 it reached even 1 mln. tons are year. Having analyzed the situation of building waste in Lithuania we can come to the conclusion mixed building waste makes the Biggest part (from 26 % to 36 %) in the flow of building waste and this shows that building waste is not sorted out on Building sites. If we want, in smaller price, to make the amounts of mixed building waste less, we should apply sorting waste gathering on the site. It means that on the building site building waste should be sorted out into separate containers and temporarily kept separately: communal, inert, suitable to recycle and reuse, secondary, dangerous and unsuitable to recycle. The people working on the building-site should be taught where and how to put different waste. In order to achieve good results workers should be taught regularly how to sort out waste in the right way. Suitable supervision of the waste holders and managers should also be applied and for this reason GPAIS was created. Now this system works irregularly and mistakes in it are quite frequent. The purpose to achieve is to create clear, equal conditions of action to all the participants of the market in order to enable them to legally process the arrangement of building and demolition waste.

Lithuania still has no regulations, it is not clear where, on what conditions recycled materials are able and allowed to be used, there is no classification of them. There are no best conveniently approachable ways of production assigned to manage building waste either in Europe or in Lithuania.

References

[1] THE EUROPEAN PARLIAMENT AND THE COUNCIL OF THE EUROPEAN UNION. On waste and repealing certain Directives. *Directive 2008/98/EC of European Parliament and of the Council of 19 November, 2008*. Available from: https://eur-lex.europa.eu/eli/dir/2008/98/oj (accessed on 17-04-2020).

[2] LIETUVOS RESPUBLIKOS APLINKOS MINISTERIJA. Dėl statybinių atliekų tvarkymo taisyklių patvirtinimo. *Lietuvos aplinkos ministro įsakymas*: 2006 m. gruodžio 29 d., Nr. D1-637 [Internet]. Available from: https://www.e-tar.lt/portal/lt/legalAct/TAR.7AB67E481C45 (accessed on 17-04-2020).

[3] Wu, Z., Yu, A. T. W., Shen, L., Liu, G. Quantifying construction and demolition waste. *Waste Management*, 2014; 34(9):1683–1692.

[4] Deloitte. *Background Paper. Workshop: Improving Management of Construction and Demolition Waste*. 25 May, 2016, Brussels.

[5] European Commission. EU construction and demolition waste management protocol. *ECORYS. European Commission Directorate – General for Internal Market, Industry, Entrepreneurship and SMEs* [Internet]. 2016; 49 p. Available from: http://ec.europa.eu/growth/content/eu-construction-and-demolition-waste-protocol-0_en (accessed on 18-04-2020).

[6] *Panevėžio regioniniame nepavojingų atliekų sąvartyne šalinamų arba į MBA / MA įrenginius priimamų mišrių komunalinių atliekų sudėties nustatymo ataskaita* [Internet]. Available from: https://pratc.lt/ataskaitos/ (accessed on 17-06-2020).

[7] *Aplinkos apsaugos agentūra* [Internet]. Available from: http://atliekos.gamta.lt/cms/index?rubricId=fbedd5df-248d-40ab-b13a-2b8034bfa270 (accessed on 18-07-2020).

Forum on Innovative Technologies and Management for Sustainability – ITMS'2020

panevėžys faculty of technologies and business

International Scientific Conference Intelligent Technologies in Logistics and Mechatronics Systems – ITELMS'2020, 1st October, 2020, Panevėžys, Lithuania

Research of Road Tanker Semi-Trailer Structure and Liquid Cargo Dynamics Interaction

Vaidas Lukoševičius[a]*, Žilvinas Bazaras[a], Vaidas Liesionis[b], Vytautas Navickas[b]

[a]*Kaunas University of Technology, Faculty of Mechanical Engineering and Design, 56 Studentu St, Kaunas LT-51424, Lithuania*
[b]*Machinery plant ASTRA, 33 Ulonų St, Alytus LT-62161, Lithuania*

Abstract

The analysis of the interaction between the liquid cargo and the tank-semi-trailer shell structure (liquid loading – Fluid Structure Interaction) was performed in the work, evaluating the analysis of situations related to changes in the tanker movement mode, various road macro profiles, dynamic load slopes and turns. Simulations of fluid sloshing in the tank with acceleration were done in different tank constructions. Dynamic analysis was performed using 1g negative acceleration, meanwhile in real life conditions tanker truck braking acceleration, using ABS (anti-lock braking system), can reach up to 0.6 g. For static and quasi-static loading cases, the analysis of a fully loaded tank truck – a semi-trailer standing on a horizontal road was performed, as well as the dynamics coefficient was estimated. It was suggested that tanker construction is strong enough but optimization further investigation is necessary.

Keywords: computational fluid dynamics; fluid-structure interaction; sloshing; liquid cargo; *Ansys.*

Introduction

Various liquid cargo transportation is often very dangerous, for example chemicals or oil products, by land roads tanker trucks are in use. Even if this choice is popular and widely used in all over the world, but tanker trucks endanger for traffic safety. As an accident result can be injured people, lost property, suffer economic loss or even death. Also in case of an accident liquid cargo can be released to environment and possibly pollute it. That is why so much attention has to be payed to prevent all of this and ensure safe liquid cargo transportation.

* Corresponding author. Tel.:+370 673 03836.
E-mail address: vaidas.lukosevicius@ktu.lt

DX20F0008

The problem of liquid sloshing in moving containers involves the evaluations of potentially destabilizing sloshing forces and moments as well as the natural frequencies of liquid free surface. Fundamentally, these properties relate to the dynamic fluid slosh and depend on several design and operating parameters in a highly nonlinear manner. These include the liquid-fill depth, liquid properties, tank geometry, nature of maneuver and designs of the ant sloshing devices, when present. Earlier attempts to study the effect of liquid sloshing on directional dynamics of tank vehicles have been performed using quasi-static. The Quasi-static method can only predict the mean steady-state position of the liquid mass center assuming a straight line (or a plane in 3D analysis) for the liquid free-surface followed by the mean steady-state responses, while it neglects the important contributions due to dynamic liquid slosh. Furthermore, the quasi-static method is not applicable to tanks with anti-slosh devices such as baffles. Although mean steady state slosh forces and moments obtained from the quasi-static analysis can be efficiently integrated to the multi-body vehicle dynamic models, crucial importance of considering the effect of dynamic liquid slosh has been emphasized in a number of studies using the computational fluid dynamics (CFD) methods These have, invariably, shown that the peak slosh forces and moments in a partially filled tank under longitudinal and / or lateral acceleration excitations are significantly greater than the mean steady-state values. Applications of CFD models for simulations of coupled fluid-vehicle systems, however, have met limited success due to not only high computational demands of the CFD methods, but also due to elaborate data transfer between the CFD and vehicle models [1–2]

The liquid sloshing in moving containers has been widely studied since 1960s. [3] presents a comprehensive review of liquid sloshing problems. However, it addresses only issues related to the space vehicle applications based on the experimental and theoretical approaches. [4–6] also presented a literature review of fluid-container interactions with focus on parameters that influence the stability of partly filled tank trucks under 6 various maneuvers. Furthermore, performed an extensive review of studies on sloshing phenomena in different applications such as liquid natural gas ship carriers, storage tanks, aerospace vehicles and road tankers. Most of the review publications in the field of liquid sloshing, however, concern aerospace and ocean vehicles as well as ground-supported structures. Fundamentally, the studies reporting fluid sloshing in road containers can be grouped in three categories based on the method of analysis: quasistatic, mechanical analogy and dynamic liquid slosh. These methods of analysis together with their essential features and limitations, in addition to methods employed for coupled fluid slosh and vehicle dynamics, are described in the following subsections [7–9].

Tankers are used to transport food or non-food liquids. Food tankers allow the transport of food liquids. These can be juices, wine, milk, oil and more. Because the products are perishable and sensitive to the environment, these products require optimal care when traveling.

In order to regulate the transport process, the European Union has set food safety standards that everyone involved in the production, transport and distribution of products must comply with. These measures include standards relating to the characteristics of the tanker. Requirements for food tankers:

- the vehicle must be equipped with a GPS system that allows the location of the truck to be tracked at any time; the material of tanker production is very important; according to EU ATP standards, it must be made of stainless steel and be isothermal;
- the food tanker must be equipped with an electric pump and using antibacterial filters;
- the temperature must be controlled by an analogue thermometer; this ensures that liquids are always stored in the best conditions;
- it is important to remember that hygiene and the cleaning process must be impeccable; that's why tank trucks are thoroughly cleaned and evaporated to avoid health hazards; different cleaning agents are used depending on the product being transported;
- food grade tankers are often separated into different compartments for the most efficient process.

The aim of this work was analysis the liquid filling tanker and the reaction of the structure to the impact load. And the weakest points finding of the tank construction at different filling levels and filling fluids.

1. Fluid-structure interaction (FSI)

Fluid-structure interaction (FSI) is the interaction of a specific moving, solid or deformable structure with the internal or surrounding fluid flow. The fluid-structure interaction can be stable or dynamic. In a dynamic interaction, a deformation caused by a fluid in a solid structure causes it to move until the source of the deformation weakens, the structure returns to its former state, and the process repeats itself. The FSI study is based on more than one physical analysis to elucidate how fluids interact with structures. Due to the flow of liquid acting on the structure, it is loaded with pressure and / or heat. These resulting loads can cause a sufficiently large deformation of the structure to change the fluid flow itself. Adverse effects on the modeled body may increase due to increased fluid-structure interactions. Fluid-structure interaction modeling is used to gain a deeper understanding of the phenomena associated with the modeled body, as well as to ensure safety, reliability and durability.

In order to maintain a high level of reliability, the software has a number of analysis methods to solve all the flui-structure interaction problems that can be encountered. *Ansys* CFDs can completely solve simple fluid-structure interaction problems. This is known as the movement of a rigid body, an example of fuel being an impeller rotating in a mixing tank.

As fluid-structure interactions increase, a more detailed assessment of the problem is required, in which case *Ansys* has an automated, easy-to-use algorithm called one-way coupling. One-way merging addresses the original CFD or *Ansys* mechanical model and automatically transfers and provides data to another system. An example of this could be the simulation of fluid flow around a conical flow meter and the automatic transmission of this data in order to calculate the resulting design response to operating factors.

To solve more complex and closely related fluid-structure interaction problems, system interconnection can be used to perform a two-way interconnected FSI simulation. Fluid and structure simulations are prepared and solved at the same time. During the calculation, the data is automatically transferred between the two solutions to obtain reliable and accurate results. An example of this could be the calculation of the flow around a rigid wing of an aircraft and the transfer of pressure loads to find out the deformations of the structure. The structural deformation is transferred back to the CFD modeling, the flow is recalculated, and this process is repeated.

Fluid-structure interactions are essential in the development of many engineering systems, e.g. aircraft, spacecraft, engines and bridges. Failure to take into account the effects of dynamic interactions can be catastrophic, especially in structures that are assembled using fatigue-sensitive materials. The Tacoma Narrows Bridge (1940) is probably one of the saddest examples of large-scale failures. Aircraft wings and turbine blades can also break due to FSI fluctuations. Fluid-structure interactions should be considered when analyzing aneurysms of large arteries and artificial heart valves. The tongues of musical instruments actually emit sound because the system of equations that regulates the dynamics of the tongue has dynamic solutions. The dynamics of leaf valves used in two-stroke engines and compressors are regulated by the FSI. The liquid-structure interaction also takes place in moving containers, where the vibrations of the liquid due to the movement of the container cause high forces and moments acting on the container structure, which have a very negative effect on the stability of the container transport system.

Fluid-structure interactions and other problems involving more than one physical phenomenon are often too complex to be solved analytically and must be analyzed experimentally or by digital modeling. Studies of computational fluid dynamics and computational structural dynamics are still ongoing, but nowadays improved simulations allow digital modeling of fluid-structure interactions. There are two main methods for modeling fluid-structure interaction problems:

Homogeneous method: the flows and design displacements on which the flow depends depend are solved simultaneously using one solution. Split method: The equations governing the flow and the displacements of the structure are solved separately by two different solutions. The homogeneous method requires code generated for this particular combination of physical problems, while the split method preserves the modularity of the software because the existing flow and structural solution are combined. In addition, the split method facilitates the solution of flow equations and structural equations by different, perhaps more efficient methods that have been developed specifically for flow or structural equations. On the other hand, the split method requires the development of a stable and accurate solution algorithm. In summary, a split solution allows you to reuse existing software, which is a nice advantage. However, the stability of the decision method needs to be taken into account. Newton-Raphson method or other fixed point iteration can be used to solve FSI problems. Methods based on the Newton-Raphson iteration are used in both the stack and split methods. These methods solve the nonlinear flow and structural equations in the whole fluid and solid region by the Newton-Raphson method. Although Newton-Raphson methods solve the problem of flow and structure state in the whole liquid and solid region, it is also possible to reformulate the FSI problem as a system in which only degrees of freedom are unknown instead of the model. This domain fragmentation concentrates the FSI error problem into a solution-related subsection. Thus, the FSI problem can be described as a root finding problem or a fixed point problem, and the solution situation is unknown [10–11].

2. Dynamic effect of liquid cargo on tanker construction

Semi-trailer tank finite element model, boundary conditions and calculation results are shown in Figures 1–4.

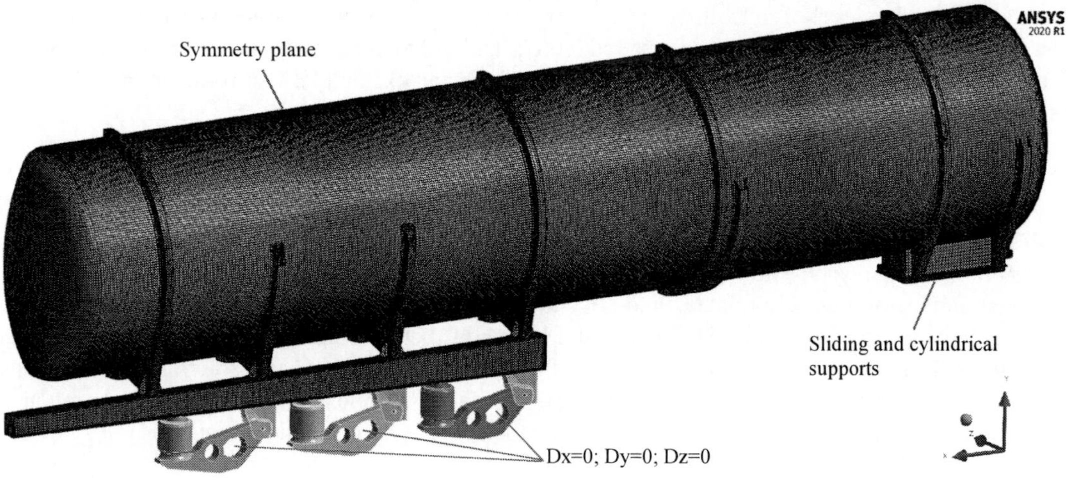

Fig. 1. Semi-trailer tank finite element model and boundary conditions

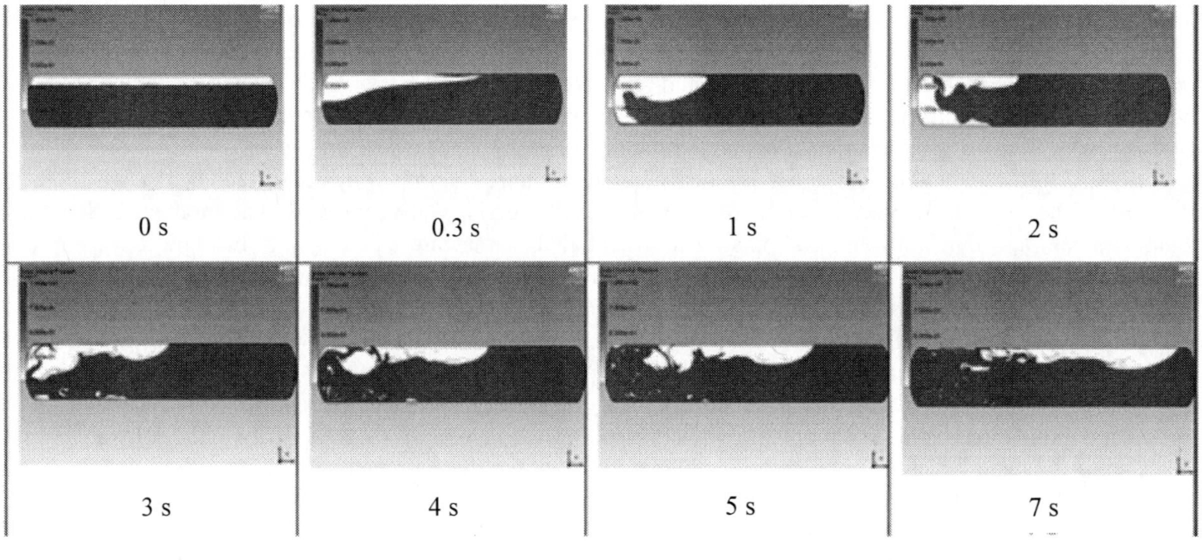

Fig. 2. Movement of liquid in a semi-trailer tank without baffles

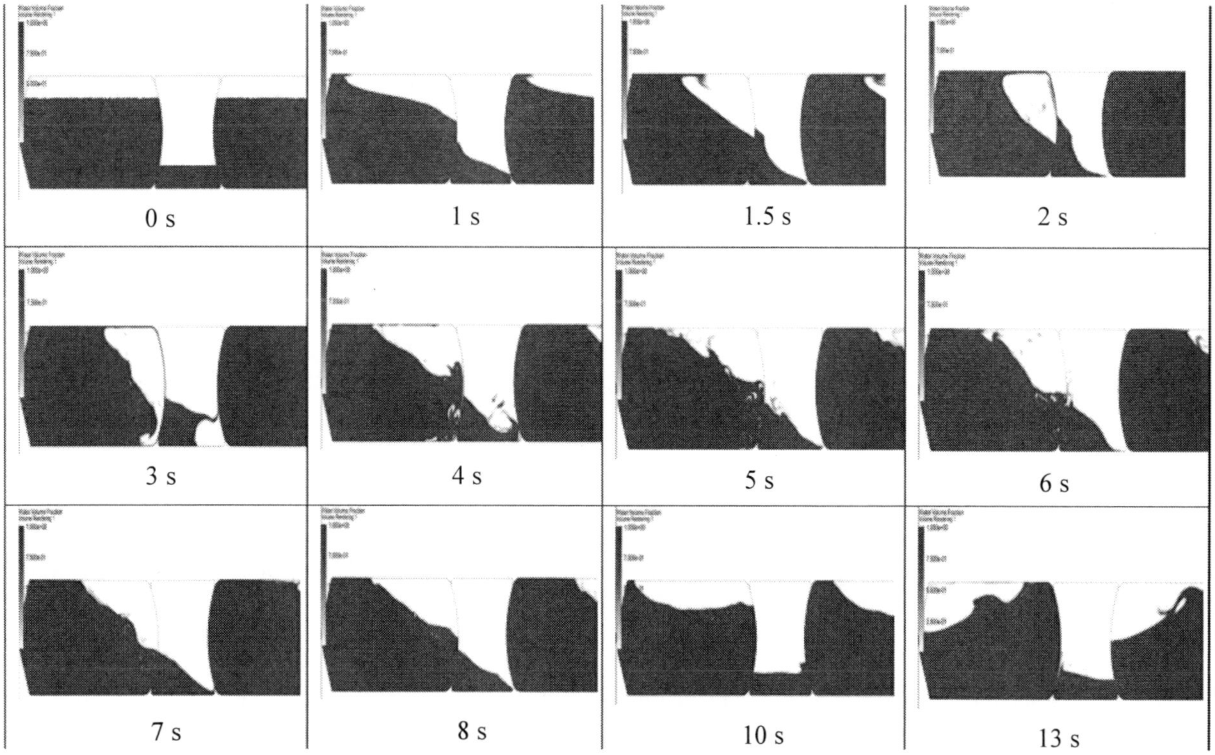

Fig. 3. Movement of liquid in a semi-trailer tank with baffles

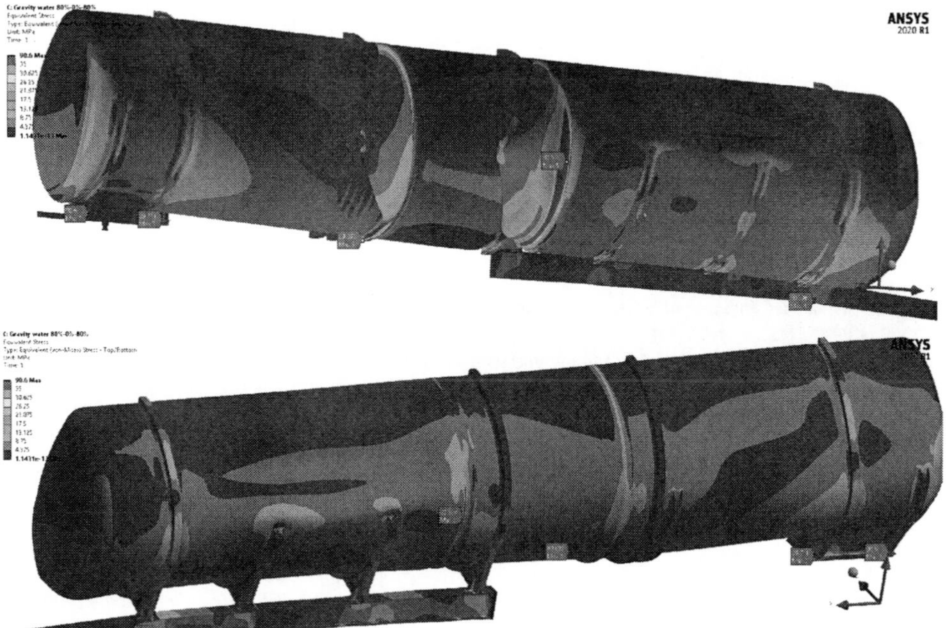

Fig. 4. Von Misses stress distribution in the semi-trailer tank construction

Calculations of the center of gravity of the construction showed that filling the tank with 80 % / 0 % / 80 % of the center of mass is obtained approximately in the center of the construction. This location allows to maintain maximum dynamic stability. After the dynamic settings of the deceleration parameters of the construction, obtained the total weight of the semi-trailer tank load of 42.485 kg. The maximum deceleration is 6 m/s^2, which is ~0.6 g. Based on the obtained results, further dynamic calculations are performed to calculate the stresses generated in the semi-trailer tank construction. The results obtained in the hydrostatic tests showed that the maximum stresses in the structure reached ~272 MPa with the syrup as the filling liquid. The most sensitive location of the structure is shown in Figure 4.

After a dynamic design test, the maximum stresses obtained are ~433 MPa with the syrup as the filling liquid. The location of the maximum stresses coincides with the location obtained during the static test. The yield strength of Duplex steel is ~450 MPa. The maximum stresses obtained in the test results are ~433 MPa. The simulation was performed with acceleration of 1 g, while in real conditions, the tanker can reach 0.6 g when braking using the ABS system. Therefore, it can be noticed that the tanker is strong enough, but further research is needed to further optimize its design.

Conclusions

The work analyzes the constructions of different types of tank trucks, their features and the reasons for the differences. The literature reviewing the prediction of free liquid surface motion is reviewed. Simulations of liquid movement in tanks of different construction were performed.

After theoretically calculating the center of mass of the train, the stopping distance and time at different speeds were also calculated.

After static tests, the maximum stresses obtained are ~272 MPa.

Dynamic tests of the liquid in the tank were performed. The maximum stresses obtained, ~433 MPa Duplex steel yield strength is ~450 MPa. The maximum stresses obtained in the test results are ~433 MPa.

The simulation was performed at an acceleration of 1 g, while in real conditions, the tanker can reach 0.6 g when braking using the ABS system. Therefore, it can be said that the tanker is strong enough, but further research is needed to further optimize its design.

Acknowledgements

This research has been supported by contract No. S-J05-LVPA-K-04-0017 "The new Duplex Steel Road Tanker Semi-Trailer development and placing on the EU market" funded by Lithuanian Business Support Agency.

References

[1] Popov, G., Sankar, S., Sankar, T. Dynamics of liquid sloshing in baffled and compartmented road containers. *Journal of Fluids and Structures*, 1993; 7(7):803–821.

[2] Armenio, V., Francescutto, A., La Rocca, M. On the roll motion of a ship with partially filled unbaffled and baffled tank (part 1): mathematical model and experimental setup. *International Journal of Offshore and Polar Engineering*, 1996; 6:278–282.

[3] Sames, P. C., Marcouly, D., Schellin, T. E. Sloshing in rectangular and cylindrical tanks. *Journal of Ship Research*, 2002; 46(3):186–200.

[4] Bauer, H. F. On the destabilizing effect of liquids in various vehicles (part 1). *Vehicle System Dynamics*, 1972; 1(3):227–260.

[5] Abramson, H. N. *The Dynamic Behavior of Liquids in Moving Containers, with Applications to Space Vehicle Technology* [Internet]. 1996. Available from: https://ntrs.nasa.gov/archive/nasa/casi.ntrs.nasa.gov/19670006555.pdf (accessed on 20-07-2020).

[6] Bauer, H. F. Fluid oscillations in the containers of a space vehicle and their influence upon stability. *NASA TR R-187*, 1964.

[7] Bogomaz, G. I., Sirota, S. A. Oscillations of a liquid in containers: methods and results of experimental studies (in Russian). *National Space Agency of Ukraine*, 2002.

[8] Celebi, M. S., Akyildiz, H. Nonlinear modeling of liquid sloshing in a moving rectangular tank. *Ocean Engineering*, 2002; 29(12):1527–1553.

[9] Rognebakke, O. F., Faltinsen, O. M. Coupling of sloshing and ship motions. *Journal of Ship Research*, 2003; 47(3):208–221.

[10] Ma, D. C., Shin, Y. S., Brochard, D., Fujita, K. Sloshing, fluid-structure interaction and structural response due to shock and impact loads. *American Society of Mechanical Engineers,* 1994. PVP-272, New York, NY (United States): e-Publishing Inc., 1999; pp. 281–304.

Forum on Innovative Technologies and Management for Sustainability – ITMS'2020

ktu 1922 panevėžys faculty of technologies and business

International Scientific Conference Intelligent Technologies in Logistics and Mechatronics Systems – ITELMS'2020, 1st October, 2020, Panevėžys, Lithuania

Gripping Devices of Industrial Robots for Manipulating Offset Dish Antenna Billets

Volodymyr Savkiv[a], Roman Mykhailyshyn[a], Pavlo Maruschak[a]*, Illia Diahovchenko[b], František Duchoň[c], Ľuboš Chovanec[c], Volodymyr Hutsaylyuk[d]

[a]*Ternopil Ivan Puluj National Technical University, Department of Automation Technological Processes and Production, 56 Ruska St, Ternopil 46001, Ukraine*
[b]*Sumy State University, Department of Electrical Power Engineering, 2 Rymskogo-Korsakova St, Sumy 40007, Ukraine*
[c]*Slovak University of Technology in Bratislava, Institute of Robotics and Cybernetics, 5 Vazovova St, Bratislava 81243, Slovak Republic*
[d]*Military University of Technology, Faculty of Mechanical Engineering, 2 Gen. S. Kaliskiego St, Warsaw 00-908, Poland*

Abstract

Invention proposes special gripping device of industrial robot for loading flat metal sheets into press and unloading of stamped plates of offset mirror antennas. The device is a T-shaped frame with three Bernoulli gripping devices pivotally mounted thereon. Design diagrams and analytical dependencies are presented for calculation of required load capacity of Bernoulli gripping devices. In the *Ansys-CFX* software environment, numerical simulation of air flow dynamics in the gap between the cooperating Bernoulli gripping device surfaces and the offset mirror antenna plate blank. The simulation was based on the Navier-Stokes average after Reynolds equations of viscous gas dynamics, the SST model of turbulence, and the γ model of laminar-turbulent transition. As a result of the simulation, the effect of the curvature radius of the surface of the plates of offset mirror antennas on the Bernoulli gripping device power characteristics was determined.

Peer-review under responsibility of the Kaunas University of Technology, Panevėžys Faculty of Technologies and Business.

Keywords: bernoulli gripping device; object manipulation; offset antenna; nozzle; radial flow; industrial robot.

* Corresponding author. Tel.: + 380 974 072696.
E-mail address: maruschak.tu.edu@gmail.com

DX20F0009

Introduction

When automating handling operations using industrial robots, the tasks of manipulating articles that change shape during machining often arise. Most often this problem is solved by cyclic automatic replacement of gripping devices, or as Shameli and Khamesee [1] use magnetic levitation, which generally reduces the productivity of the technological operation.

In this work there is proposed a special gripping device capable of gripping a flat metal sheet for loading into a press, and after it is pressed to grip a blank of a plate of an offset mirror antenna, has acquired a parabolic shape. The main elements of the proposed special gripping device are three Bernoulli grips pivotally mounted on a T-shaped frame. Bernoulli gripping devices (BGD) have a number of advantages: Erzincanli, Ozcelik [2–3] did research for transportation of non-rigid objects, for transportation of food in production and finished products Davis et al. [4] and Petterson et al. [5], for transportation of skin Dini et al. [6].

Li et al. in paper [7] the pressure distribution and suction force are measured experimentally. A theoretical model of the air flow between the gripper and the workpiece is created, based on which the theoretical formulas for calculating the pressure distribution and suction force are derived by Li et al. [8]. It is found that the outer diameter of the gripper has a major impact on the suction force, and its design is closely related to the gap height and the supply mass flow rate. Then, the relationship between the outer diameter and the suction force and that between the gap height and the suction force are discussed, based on which a method for finding the optimal outer diameter is presented. In paper [9] Li et al. study experimentally and theoretically investigates the dynamic characteristics of the BG.

It is most of all investigated and introduced on production BGD with cylindrical or circular nozzle and vortex grippers. For the purpose of minimization of energy consumption of BGD when performing handling operations by authors of the article the method of optimization of gripper orientation in the course of manipulation was developed. The method of optimization of BGD orientation when performing transport operations on a rectilinear and arc trajectory is provided Savkiv et al. in the articles [10–11]. Influence of force of front resistance of Q_1, Q_2 on the minimum necessary lifting force is investigated Mykhailyshyn et al. in article [12]. The description of experimental installation and the analysis of the received experimental results on application of a method of optimization of BGD orientation is described Mykhailyshyn et al. in the paper [13]. Gasdynamic analysis of the BG interaction with the surface of flat objects with displacement of the center of mass Savkiv, Mykhailyshyn et al. carried out in [14–15]. Also authors in the article Savkiv, Mykhailyshyn et al. [16–18] deals with the topical issue of reducing energy consumption for transportation of industrial objects. The energy efficiency of the process of OM with the use of the orientation optimization method while gripping with the help of different methods has been studied. The economic efficiency of the use of the optimal orientation of BGD while transporting the OM in comparison to the transportation without re-orientation has been proved. Influence of parameters of a gripping system on power expenses of the industrial robot during transportation is investigated Mykhailyshyn et al. in article [19]. However, Wagner at al. in [20] condition and dynamics of course of gas stream in snivel and in radial interval between the interacting surfaces of BGD and OM, methods of increase in lifting ability of these devices due to optimization of their design data are insufficiently studied and demand further research.

Nomenclature

r_n	radius of nozzle;
h_c	the distance between the interacting surfaces of the manipulation object and the gripping device;
p_a	air pressure;
N_1, N_2	reaction acting at tab contact points with antenna blank surface;
mg	weight of offset mirror antenna blank;
α	angle between vertical and axis of gripping device;
F_s	spring elastic force;
k	coefficient of spring elasticity;
a	distance from gripper axis to spring axis;
b	distance from BGD pivot axis to its center of mass;
F_l	lifting force of gripping device.

1. Structure and operating principle of gripping devices

The structure and sections of the individual mechanisms GD of the containment for the offset dish antenna plate are shown in Figure 1. Gripping device consists of three similar BGD 1, 2, 3 (Fig. 1, a–c) hinged on T-shaped frame 4. BGD have the possibility of mutual horizontal displacement by means of pins 5 rigidly connected to their upper part, which are aligned with horizontal slots 6 (Fig. 1, d) of T-shaped frame.

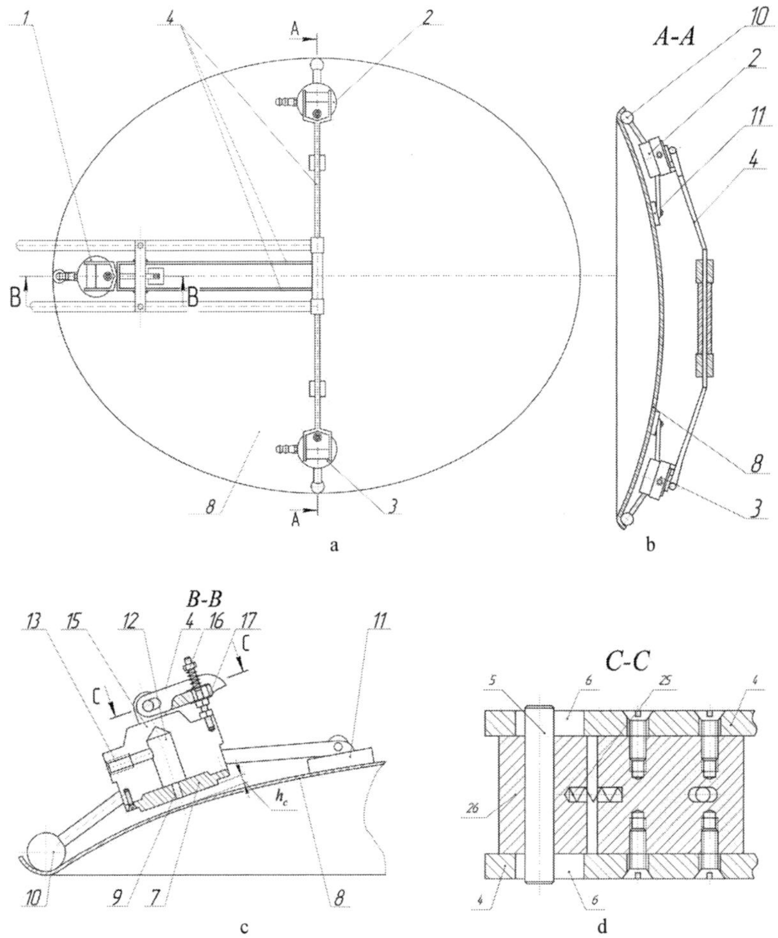

Fig. 1. Design and cross-sections of separate mechanisms of gripping devices for plates of offset dish antennas

Disk 7 is mounted in each BGD, in axial part of which nozzle 9 is made along normal line to object of processing 8. Two performances 10 and 11 are arranged on different sides of disc 7. The first step 10 in each BGD is arranged around the periphery of the processing object 8, and the second step 11 of each BGD is arranged in series towards the center of the processed object 8. The axis of each pin 5 is perpendicular to the plane passing through the axis of the respective nozzle and the center of the processing object 8.

In the initial state (until the first step 10 enters the radius slot of the blank offset dish antenna 8) with compression spring 14 between the body 15 of the Bernoulli gripping device and T-shaped frame 4, the axle 5 together with the body 15 is pressed into the extreme left position of the slot 6 (Fig. 1, c–d). However, the compression spring 16 comprises the housing 15 in its extreme, counterclockwise inclined position defined by the stop 17.

After forming the blank of the offset dish antenna 8 in the mold, the industrial robot introduces a T-shaped frame 4 into the space above the blank. Positioning of the frame is performed so that the first performance 10 of each of the Bernoulli grippers is set to the lower point of the radial slot of the antenna blank. The robot then performs a further downward shift of the frame, in which each of the Bernoulli grippers is returned to a corresponding angle until contact of the second projection 11 with the antenna blank is reached. At the same time axis 5, is not movably established in building 15 is displaced concerning a groove 6 to the right and down, the spring 16 contracts, and the spring 14 is unclenched at necessary sizes, providing continuity of a kinematic chain from performances 10, 11 to T-shaped frames 4.

After all Bernoulli grippers are installed in the specified position, compressed air is supplied to chamber 12 through channel 13 (Fig. 1, c). Air flows through nozzle 9 into clearance between lower part of disk 7 and surface of antenna blank. At that, at radius equal to nozzle radius r_n, at values of distance between interacting surfaces $h_c < r_n / 2$, flow experiences greatest narrowing. At the point of greatest narrowing, at excessive supply pressures of the pickup by compressed air more than 30.0 kPa, the flow reaches the critical speed equal to the speed of sound for these conditions. As a result of further increase of radial flow area, its supersonic speed increases, and static pressure on OM surface decreases to value of lower atmospheric value. At some distance from the nozzle center there is sharp braking of supersonic flow, with its subsequent transition to subsonic flow, which is accompanied by formation of pressure jump. As a result of further expansion, the subsonic flow rate drops and the static pressure in the gap smoothly increases to atmospheric p_a. The effect of the vacuum on the surface of the antenna blank results in a lifting force towards the end portion of the disc 7.

2. Analysis of the power characteristics of the gripping device

As can be seen from Figure 1 (a, c) in static mode, the offset dish antenna blank is sufficient to be held due to the force of only BGD 2 and 3. At the same time BGD 1 is necessary to ensure location of offset mirror antenna blank at its gripping, as well as to prevent antenna rocking at acceleration (braking) of final link operation.

Refer to Figure 2 for design diagrams for determining the required lifting force of Bernoulli grippers.

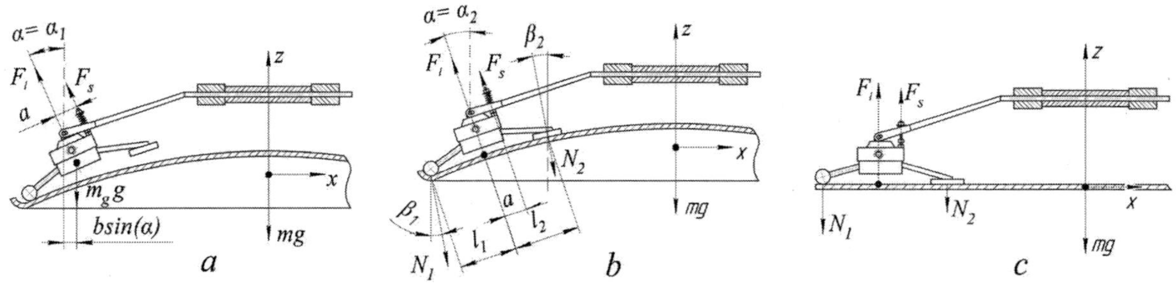

Fig. 2. Diagrams for determination of required load capacity of BGD (a – positioning of frame with gripping devices above offset antenna blank; b – grip of offset antenna blank; c – gripping of flat blank)

The minimum load capacity of the BGD 2 and 3 required to contain the stamped antenna blank can be determined from the equilibrium condition of all forces (Fig. 2, b) acting on this blank at $N_2 = 0$. By designing all forces on the z-axis, we define the condition of antenna blank content:

$$2F_l \cos\alpha > 2N_1 \cos\beta_1 + mg; \tag{1}$$

where N_1 is the reaction, acting at the point of contact of the tab 10 with the surface of the antenna blank, β_1 is the angle between the vertical and the normal to the tangent to the parabolic surface of the antenna blank at the point of contact of the tab 10 with its surface.

The reaction N_1 can be determined from the BGD equilibrium condition. Taking into account the condition of equilibrium of moments of forces relative to the axis of rotation of the gripping device and the specified condition $N_2 = 0$, we will find:

$$N_1 = \frac{F_s a - m_g g b \sin\alpha}{l_1 \cos(\alpha - \beta_1)}; \tag{2}$$

where F_s is the elastic force of the spring 16; l_1 is the distance from the grip axis to the center of the tab 10.

Elastic force of spring 16:

$$F_s = k\left(\delta_1 + a(\alpha_1 - \alpha)\frac{\pi}{180^\circ}\right); \tag{3}$$

where δ_1 is the value of preliminary compression of spring 16, which provides equilibrium state of BGD at frame positioning above offset antenna blank (Fig. 2, a) α_1 – angle of inclination of BGD axis, which corresponds to its equilibrium state in initial position.

The initial angle of inclination of the BGD axis is set by the previous compression of the spring 16 so that $\alpha_1 = \alpha_2 + (5...7)^\circ$. Angle value α_2 depends on geometric parameters of offset antenna blank and design parameters of gripping system. The relationship between the pre-compression value of the spring 16 and its coefficient of elasticity can be determined from formulas (2) and (3), taking into account when positioning the BGD frame above the offset antenna blank (Fig. 2, a) N_1 and N_2 are zero. Then:

$$\delta_1 = \frac{m_g g b \sin\alpha_1}{ka}. \tag{4}$$

On the basis of formulas (1–3), find the required load capacity of BGD 2 and 3 by substituting $\alpha = \alpha_2$:

$$F_l > \frac{1}{\cos\alpha_2}\left(\frac{ka\left(\delta_1 + a(\alpha_1 - \alpha_2)\frac{\pi}{180^\circ}\right) - m_g g b \sin\alpha_2}{l_1 \cos(\alpha_2 - \beta_1)} + \frac{mg}{2}\right). \tag{5}$$

Find the required load capacity of BGD 2 and 3 for flat blank content (Fig. 2, c) by substitution in formula (5) $\alpha = 0$, $\beta_1 = 0$:

$$F_l > \frac{ka}{l_1}\left(\delta_1 + a\alpha_1 \frac{\pi}{180^\circ}\right) + \frac{mg}{2}. \tag{6}$$

In general, the lifting force by gripping the workpiece can be determined by integrating the absolute pressure distribution p_r between their cooperating surfaces:

$$F_l = 2\pi \int_0^{r_g} (p_a - p_r)\, r dr. \tag{7}$$

Using the technique described in [21], it is possible to determine the pressure distribution in the radial space for the case of interaction of the Bernoulli gripping device with the flat surface of the transport object. However, it is quite difficult to analytically evaluate the power interaction of the BGD with the fragment of the parabolic surface of the offset antenna. For this purpose, it is better to use numerical simulation of air flow dynamics in the chamber, nozzles of BGD and in the interval between its flat surface and parabolic surface of antenna blank.

To simulate the air flow between the active surfaces of the Bernoulli gripping device and the antenna blank, we will use Reynolds-averaged Navier-Stokes equations (RANS) Snegiryov, Garbaruk et al. [22, 23]. At the same time, the influence of mass forces is neglected. Then the system of basic equations will be as follows:
equation of continuity of a stream:

$$\frac{\partial \rho}{\partial t} + \frac{\partial(\rho \cdot V_j)}{\partial x_j} = 0; \tag{8}$$

impulse equation:

$$\rho \frac{\partial V_i}{\partial t} + \rho \cdot V_j \cdot \frac{\partial V_i}{\partial x_j} = \frac{\partial \tau_{ij}}{\partial x_j}; \tag{9}$$

energy equation:

$$\rho \frac{\partial E}{\partial t} + \rho \cdot V_j \cdot \frac{\partial E}{\partial x_j} = -\frac{\partial q_j}{\partial x_j} + \frac{\partial}{\partial x_j}\left(\tau_{ij} \cdot V_i\right); \tag{10}$$

ideal gas state equation:

$$\rho = \frac{p}{R \cdot T}; \tag{11}$$

where i, j – indices, take values of 1, 2, 3; ρ – air density; t – time; x – coordinate; V – vector of air velocity; τ_{ij} – stress tensor; E is the total energy of air; q is a heat flux density vector taking into account heat transfer by heat conduction and diffusion; R is gas constant; T is the absolute air temperature.

The system of equations (8–11) must be supplemented by a turbulence model. Shear stress transfer model (SST) is selected to describe turbulence Menter et al. [24]. The SST model of turbulence is supplemented by transition models Menter et al. [25, 26], which allow to predict the position of the laminar-turbulent transition. These models are based on the interleavability coefficient γ, whose value at a given point is determined by the ratio of the flow of which a turbulent mode is observed to the total observation time.

Sufficient accuracy of calculations is given by the laminar-turbulent transition model with another differential equation for the value γ Menter et al. [27]:

$$\frac{\partial(p\gamma)}{\partial t} + \frac{\partial(pV_j\gamma)}{\partial x_j} = P_\gamma - E_\gamma + \frac{\partial}{\partial x_j}\left[\left(\mu + \frac{\mu_t}{\sigma_\gamma}\right)\frac{\partial \gamma}{\partial x_j}\right]; \tag{12}$$

where P_γ, E_γ – respectively generative and dissipation members of managing directors of laminar and turbulent transition; μ – molecular dynamic viscosity of gas; μ_t – turbulent dynamic viscosity of gas; $\sigma_\gamma = 1.0$ – model constant.

The modified equations of the SST turbulence for this γ – model laminar-turbulent transition are as follows Menter et al. [24]:

$$\frac{\partial}{\partial t}(\rho k) + \frac{\partial}{\partial x_j}(\rho V_j k) = \tilde{P}_k + P_k^{\lim} - \tilde{D}_k + \frac{\partial}{\partial x_j}\left((\mu + \sigma_k \mu_t)\frac{\partial k}{\partial x_j}\right); \tag{13}$$

$$\frac{\partial}{\partial t}(\rho \omega) + \frac{\partial}{\partial x_j}(\rho V_j \omega) = \alpha \frac{P_k}{\nu_t} - D_\omega + Cd_\omega + \frac{\partial}{\partial x_j}\left((\mu + \sigma_\omega \mu_t)\frac{\partial \omega}{\partial x_j}\right); \tag{14}$$

where k – kinetic turbulent energy; ω – the specific speed of dissipation of kinetic energy of turbulence; P_k, D_k – original generation and dissipation of the SST model; $P_k^{\lim}$ – the additional part, which provides the correct gain of turbulent viscosity in transitional area at very low level of turbulent viscosity of the running stream; ν_t – turbulent kinematic viscosity of gas; σ_k, α, a_1 – empirical constants of model.

3. Results and discussion

The force of attraction by the Bernoulli gripping device of the object of transportation is influenced by the following factors: pressure of supply of the gripping device with compressed air; geometrical parameters of a nozzle; geometric parameters of the surface of the transport object with which the gripping device interacts.

In order to assess the effect of curvature radii (convexities) of fragments of parabolic surface of offset antenna blank on force characteristics of Bernoulli gripping devices interacting with these fragments, it is possible to replace parabolic surface with equivalent spherical surface. Such change gives a slight error, since within the fragment of the parabolic surface with which the Bernoulli gripping device interacts, the radius of curvature of this surface varies by less than 4 %. The radius of curvature of the spherical surface shall be changed within 0.6...3 m, which corresponds to the dimensions of offset antennas 0.85...2.15 m.

Numerical simulation of the dynamics of air flow in the chamber, nozzles of the BGD and in the interval between its flat surface and the spherical surface of the object of transportation was performed in the environment of computational fluid and gas dynamics *Ansys-CFX* using RANS and γ – model turbulence. For modeling in this software environment, an unstructured finite difference grid is built in the simulation area. The total number of nodes in the design area is 2.4–3.2 million. Mesh nodes are combined into three-dimensional elements (tetrahedra and prisms). The total number of volume elements of the grid is 4.6–6.3 million. The total number of tetrahedra is 2.3–3.4 million. In the simulation, it was set: air is the ideal gas; thermodynamic process is adiabatic. The boundary conditions for the airflow model are shown in Figure 3. The simulation was carried out for BGD with nozzle radius r_n = 3 mm and outer radius r_g = 30 mm.

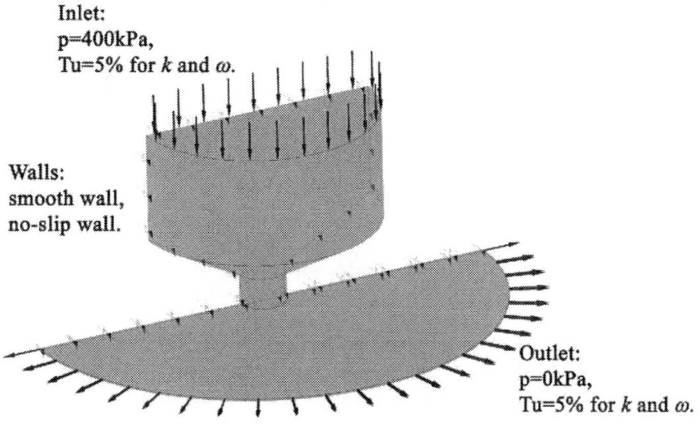

Fig. 3. Air flow model limits

Based on the results of the simulation using the decisive sonicTurbFoam module (for turbulent flows of compressible gases moving at sound and supersonic speeds), graphs of pressure distribution in the interval between the flat end of the BGD and the spherical surface were built for different values of radii R of this surface at the distance from the nozzle to the spherical surface h_c = 0.2 mm (Fig. 4).

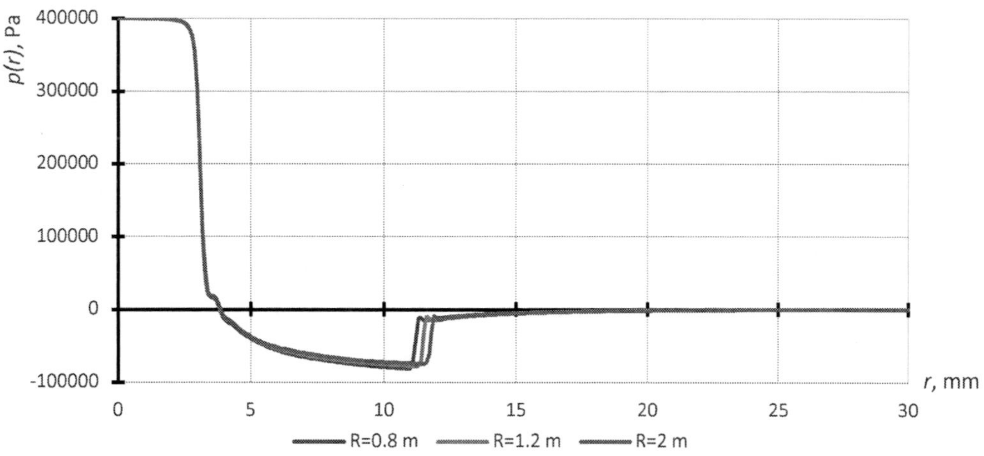

Fig. 4. Plots of pressure distribution in the interval between the flat end of the BGD and the spherical surface for different values of radii R of this surface

The graphs in Figure 4 show that as the curvature radius of the surface of the manipulation object increases, the size of the supersonic vacuum zone on that surface increases, but the amount of vacuum itself decreases. The radius of curvature of the surface of the manipulation object has minimal influence on the value of the vacuum value in the subsonic zone.

In order to determine the effect of the curvature radius of the surface of the manipulation object on the carrying capacity of the Bernoulli gripping, numerical integration of the pressure distribution data in the radial gap according to formula (7). As a result, it is revealed that at the radii of curvature of the surface of the object of manipulation more than 3 m the lifting capacity of the BGD reaches the maximum value corresponding to the lifting force of flat objects. It has also been found that when gripping an offset antenna blank with dimensions of 0.85–1.05 m (the minimum radius of curvature of the parabolic surface is 0.6 m), the load capacity of the Bernoulli gripping device will decrease by 39 % compared to the lifting force of the flat object. The results of the calculations for the two values of the distance between the BGD and the spherical surface are shown in Figure 5.

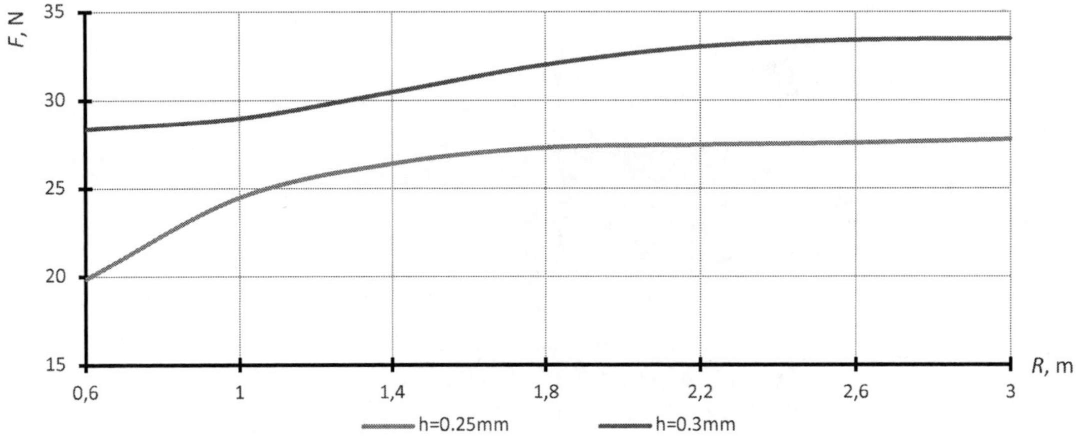

Fig. 5. Dependence of the load capacity of the Bernoulli gripping device on the radius of curvature of the surface of the content of the transportation object

Analysis of these graphs shows that as the radius of curvature of the surface of the object of manipulation increases from 0.6 to 1.8 m, the lifting force of this object by Bernoulli gripper increases by 13...37 %, and if the radius increases from 1.8 to 3.0 m only by 2...4 % .

Thus, after determining the required load capacity of BGD 2 and 3 by formula (5), appropriate correction is made for increase of their power characteristics depending on diameter of offset antenna blank. The lifting capacity of the BGD is generally increased by increasing the supply pressure.

By means of additional rotation of the BGD, if there are changes in the shape of the surface of the object to be gripped, it is possible to provide the necessary gap between the lower surface of the disc 7 and the surface of the object, at which the lifting force of each of the grippers would be maximum.

The economic effect of the implementation of this device is ensured by the possibility to use it for different processing objects in different form, allows to increase the productivity of the technological process.

Conclusions

Bernoulli gripping devices have been proven to allow both flat and objects with convex surfaces to be contained. This allows them to be effectively used as elements of special gripping devices of industrial robots for loading flat metal sheets into the press and unloading stamped plates of offset mirror antennas of parabolic shape.

As a result of the numerical simulation of the dynamics of the air flow in the gap between the interacting Bernoulli gripping device surfaces and the plate blank of the offset mirror antenna, it has been found that as the radius of curvature of the surface of the manipulation object increases, the size of the supersonic vacuum zone on this surface increases, but the amount of vacuum itself decreases. It has also been found that the value of the vacuum value in the subsonic zone has a minimal effect on the radius of curvature of the surface of the manipulation object.

With the curvature radii of more than 3 m of manipulation object surfaces, the Bernoulli gripping device load capacity reaches the maximum value and approaches the value of the lifting force of the flat objects. It has also been found that when gripping a blank of an offset antenna with dimensions of 0.85–1.05 m, the load capacity of the Bernoulli gripping device will decrease by 39 % compared to the lifting force of the flat object.

References

[1] Shameli, E., Khamesee, M. B., Huissoon, J. P. Nonlinear controller design for a magnetic levitation device. *Microsystem Technologies*, 2007; 13(8):831–835.

[2] Toklu, E., Erzincanli, F. Modeling of radial flow on a non-contact end effector for robotic handling of non-rigid material. *Journal of Applied Research and Technology*, 2012; 10(4):590–596.

[3] Ozcelik, B., Erzincanli, F. Examination of the movement of a woven fabric in the horizontal direction using a non-contact end-effector. *The International Journal of Advanced Manufacturing Technology*, 2005; 25:527–532.

[4] Davis, S., Gray, J., Caldwell, D. An end effector based on the Bernoulli principle for handling sliced fruit and vegetables. *Robotics and Computer-Integrated Manufacturing*, 2008; 24(2):249–257.

[5] Petterson, A., Ohlsson, T., Caldwell, D. G., Davis, S., Gray, J. O., Dodd, T. J. A Bernoulli principle gripper for handling of planar and 3D (food) products. *Industrial Robot: An International Journal*, 2010; 37(6):518–526.

[6] Dini, G., Fantoni, G., Failli, F. Grasping leather plies by Bernoulli grippers. *CIRP Annals*, 2009; 58(1):21–24.

[7] Li, X., Kagawa, T. Theoretical and experimental study of factors affecting the suction force of a Bernoulli gripper. *Journal of Engineering Mechanics*, 2014; 140(9):04014066.

[8] Shi, K., Li, X. Optimization of outer diameter of Bernoulli gripper. *Experimental Thermal and Fluid Science*, 2016; 77:284–294.

[9] Shi, K., Li, X. Experimental and theoretical study of dynamic characteristics of Bernoulli gripper. *Precision Engineering*, 2018; 52:323–331.

[10] Savkiv, V., Mykhailyshyn, R., Fendo, O., Mykhailyshyn, M. Orientation modeling of Bernoulli gripper device with off-centered masses of the manipulating object. *Procedia Engineering*, 2017; 187:264–271.

[11] Savkiv, V., Mykhailyshyn, R., Duchon, F., Mikhailishin, M. Modeling of Bernoulli gripping device orientation when manipulating objects along the arc. *International Journal of Advanced Robotic Systems*, 2018; 15(2):1729881418762670.

[12] Mykhailyshyn, R., Savkiv, V., Duchon, F., Koloskov, V., Diahovchenko, I. Analysis of frontal resistance force influence during manipulation of dimensional objects. *2018 IEEE 3rd International Conference on Intelligent Energy and Power Systems (IEPS)*, 2018; pp. 301–305.

[13] Mykhailyshyn, R., Savkiv, V., Mikhalishin, M., Duchon, F. Experimental research of the manipulation process by the objects using Bernoulli gripping devices. *Young Scientists Forum on Applied Physics and Engineering*, 2017; pp. 8–11.

[14] Mykhailyshyn, R., Savkiv, V., Duchon, F., Maruschak, P., Prentkovskis, O. Substantiation of Bernoulli grippers parameters at non-contact transportation of objects with a displaced center of mass. *Transport Means – Proceedings of the International Conference*, 2018; pp. 1370–1375.

[15] Maruschak, P., Savkiv, V., Mykhailyshyn, R., Duchon, F., Chovanec, L. The analysis of influence of a nozzle form of the Bernoulli gripping devices on its energy efficiency. *Proceedings of ICCPT* 2019, TNTU, Scientific Publishing House "SciView", 2019; pp. 66–74.

[16] Savkiv, V., Mykhailyshyn, R., Maruschak, P., Chovanec, L., Prada, E., Virgala, I., Prentkovskis, O. Optimization of design parameters of Bernoulli gripper with an annular nozzle. *Transport Means – Proceedings of the International Conference*, 2019; pp. 423–428.

[17] Savkiv, V., Mykhailyshyn, R., Duchon, F., Mikhalishin, M. Energy efficiency analysis of the manipulation process by the industrial objects with the use of Bernoulli gripping devices. *Journal of Electrical Engineering*, 2017; 68 (6):496–502.

[18] Mykhailyshyn, R., Savkiv, V., Duchon, F., Trembach, R., Diahovchenko, I. M. Research of energy efficiency of manipulation of dimensional objects with the use of pneumatic gripping devices. *IEEE 2nd Ukraine Conference on Electrical and Computer Engineering (UKRCON)*, 2019; pp. 527–532.

[19] Mykhailyshyn, R., Savkiv, V., Duchon, F., Koloskov, V., Diahovchenko, I. Investigation of the energy consumption on performance of handling operations taking into account parameters of the grasping system. *IEEE 3rd International Conference on Intelligent Energy and Power Systems (IEPS)*, 2018; pp. 295–300.

[20] Wagner, M., Chen, X., Nayyerloo, M., Wang, W., Chase, J. G. A novel wall climbing robot based on Bernoulli effect. *IEEE/ASME International Conference on Mechtronic and Embedded Systems and Applications*, 2008; pp. 210–215.

[21] Savkiv, V., Mykhailyshyn, R., Duchon, F. Gasdynamic analysis of the Bernoulli grippers interaction with the surface of flat objects with displacement of the center of mass. *Vacuum*, 2019; 159:524–533.

[22] Snegiryov, A. Y. High-performance computing in technical physics. *Numerical Simulation of Turbulent Flows*. S. Petersburg, Polytechnic University Publ., 2009.

[23] Garbaruk, A. V. Modern approaches to modeling turbulence. *S. Petersburg, Polytechnic University Publ.* 2016.

[24] Menter, F. R. Two-equation eddy-viscosity turbulence models for engineering applications. *AIAA Journal* 1994; 32(8): 1598–1605.

[25] Menter, F. R., Esch, T., Kubacki, S. Transition modelling based on local variables. *Engineering Turbulence Modelling and Experiments* 2002; pp. 555–564.

[26] Menter, F. R., Langtry, R. B., Völker, S. Transition modelling for general purpose CFD codes. *Journal Flow Turbulence and Combustion* 2006; 77:277–303.

[27] Menter, F. R., Smirnov, P. E., Liu, T., Avancha, R. A one-eoequation local correlation-based transition model. *Flow Turbulence Combust* 2015; s10494–015–9622–4.

Forum on Innovative Technologies and Management for Sustainability – ITMS'2020

International Scientific Conference Intelligent Technologies in Logistics and Mechatronics Systems – ITELMS'2020, 1st October, 2020, Panevėžys, Lithuania

Determination of Stress Strain State in Multilayer Cylinder Subjected to Hydrostatic Pressure under Elastic Loading

Dainius Vaičiulis[a]*

[a]*Kaunas University of Technology, Panevėžys Faculty of Technologies and Business, 33 Nemuno St, Panevėžys LT-37164, Lithuania*

Abstract

This paper presents expressions for determining the stress strain state in a multilayer cylinder (pipe) subjected to hydrostatic pressure under elastic loading. The accuracy of presented method was verified by the finite element method. A good agreement between the analytical and numerical results was obtained.

Peer-review under responsibility of the Kaunas University of Technology, Panevėžys Faculty of Technologies and Business.

Keywords: multilayer cylinder; stress strain state; elastic loading; pressure; layer interference.

Introduction

Multilayer pipes can consist of many layers, with different functions. Multilayer pipes are widely used in industry and household since they have many potential advantages over pipes made from typical homogeneous materials, such as specific stiffness strength, good corrosion resistance, light weight and good thermal properties [1]. Most gas or liquid transportation pipelines have cylindrical shape.

Investigation of multilayer pipes usually carried out analytically, experimentally or by using FEA [2–5].

Although the principles of determining the stress state in a multilayer cylinder are known for a long time [6–9] but no works has been found which present an analytical solution of stress strain state in multilayer cylinder subjected to hydrostatic-pressure.

The analytical method for stress strain state determination in n^{th} layer cylinder under elastic loading in this paper is presented.

* Corresponding author.
E-mail address: Dainius.Vaiciulis@ktu.lt

DX20F0010

Nomenclature

$A_{in;j}$	parameter at inner radius of jth cylinder layer;
$A_{out;j}$	parameter at outer radius of jth cylinder layer;
a_j	first parameter of jth cylinder layer material depends on stress strain state;
$B_{j;j+1}$	parameter combining j^{th} and $(j+1)^{th}$ cylinder layers parameters;
b_j	second parameter of jth cylinder layer material depends on stress strain state;
E	elastic (*Young*'s) modulus of any cylinder layer;
E_j	elastic modulus of jth cylinder layer;
e	strain without indicating strain state component;
e_i	strain intensity (*von Mises* criteria) of any cylinder layer;
e_r	radial strain of any cylinder layer;
e_z	axial strain of any cylinder layer;
e_θ	hoop strain of any cylinder layer;
j	index of multilayer cylinder layer;
n	count of cylinder layers;
r	any valid radius of cylinder;
r_j	outer radius of jth cylinder layer;
p_0	internal pressure of cylinder;
p_j	contact pressure between j^{th} and $(j+1)^{th}$ cylinder layers ($j > 0$ and $j < n$);
p_n	external pressure of cylinder;
u	radial displacement of any cylinder layer;
$u_j(r)$	radial displacement of jth cylinder layer at radius r;
v	Poisson's ratio of any cylinder layer;
v_j	Poisson's ratio of jth cylinder layer;
$\Delta_{j;j+1}$	interference between j^{th} and $(j+1)^{th}$ cylinder layers;
σ	stress without indicating stress state component;
σ_i	stress intensity (*von Mises* criteria) of any cylinder layer;
σ_r	radial stress of any cylinder layer;
$\sigma_{r;j}(r)$	radial stress of jth cylinder layer at radius r;
σ_z	axial stress of any cylinder layer respectively;
σ_θ	hoop stress of any cylinder layer;
$\sigma_{\theta,j}(r)$	hoop stress of jth cylinder layer at radius r;
σ_y	yield stress of any cylinder layer.

1. Determination of stress strain state

Calculation scheme of multilayer cylinder subjected to hydrostatic pressure is presented in Figure 1.

The stress strain state is determined by using these assumptions: all pipe materials are isotropic and linearly elastic; an internal or (and) external pressure are the only loads; multilayer pipe does not loose stability.

Internal multilayer cylinder layer index is 1, external – n. Radial and hoop stresses in j^{th} layer at radius r may be determined by *Lame*'s equations [6]:

$$\left.\begin{matrix}\sigma_{r;j}(r)\\ \sigma_{\theta;j}(r)\end{matrix}\right\} = \frac{p_{j-1}\,r_{j-1}^2}{r_j^2 - r_{j-1}^2}\left(1 \mp \frac{r_j^2}{r^2}\right) - \frac{p_j\,r_j^2}{r_j^2 - r_{j-1}^2}\left(1 \mp \frac{r_{j-1}^2}{r^2}\right). \tag{1}$$

Axial stress value depends on stress strain state. For axisymmetric plane stress state:

$$\sigma_z = 0. \tag{2}$$

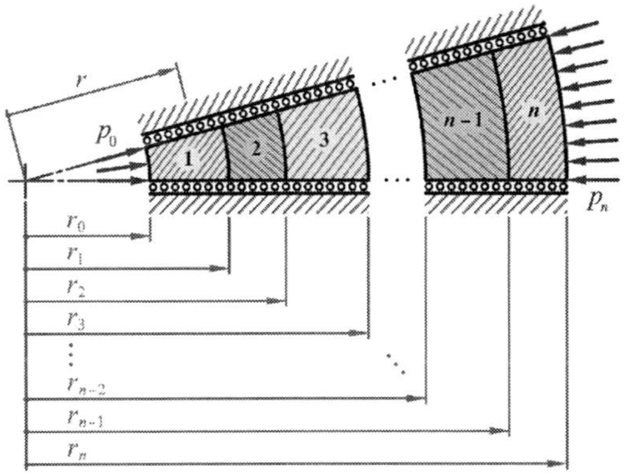

Fig. 1. Calculation scheme of multilayer cylinder subjected to hydrostatic pressure

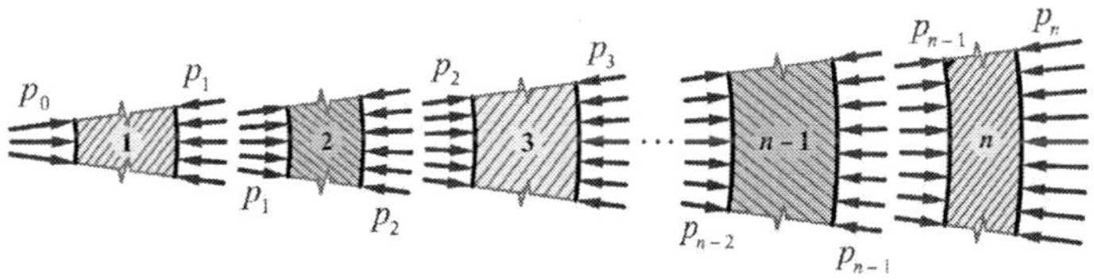

Fig. 2. The pressures acting on each layer of the multilayer cylinder in the general case

For axisymmetric plane strain state:

$$\sigma_z = \nu\left(\sigma_r + \sigma_\theta\right). \tag{3}$$

In order to calculate the stress state components additionally the contact pressures p_1, p_2, …, p_{n-1} must be known (Fig. 2). They will be calculated from the conditions that at the contact of any two layers the sum of radial displacement is equal to interference of these layers.

Radial displacement can be expressed from hoop strain $e_\theta = u / r$ and the generalized *Hooke*'s law $e_\theta = [\sigma_\theta - \nu\,(\sigma_r + \sigma_z)] / E$:

$$u = \frac{r}{E}\left[\sigma_\theta - \nu\left(\sigma_r + \sigma_z\right)\right]. \tag{4}$$

As mentioned above the sum of the radial displacements of any two layers of multilayer cylinder at their contact surface must be equal to interference between these layers, i.e.:

$$u_{j+1}\left(r_j\right) - u_j\left(r_j\right) = \Delta_{j;\,j+1}; \tag{5}$$

where the "–" sign is used because the displacement of the contact surface of the layer closer to the axis of the cylinder due to interference is always negative.

Condition (5) for every contact surface between j^{th} and $(j + 1)^{th}$ layers by evaluating Eqs. (1)–(4):

$$A_{in;\,j}\, p_{j-1} + B_{j;\,j+1}\, p_j + A_{out;\,j+1}\, p_{j+1} = \frac{\Delta_{j;\,j+1}}{r_j}; \tag{6}$$

where $A_{in;\,j} = -\dfrac{2\,a_j\, r_{j-1}^2}{E_j\left(r_j^2 - r_{j-1}^2\right)}$; $A_{out;\,j} = -\dfrac{2\,a_j\, r_j^2}{E_j\left(r_j^2 - r_{j-1}^2\right)}$; $B_{j;\,j+1} = \dfrac{a_j\left(r_j^2 + r_{j-1}^2\right)}{E_j\left(r_j^2 - r_{j-1}^2\right)} - \dfrac{b_j\,\nu_j}{E_j} + \dfrac{a_{j+1}\left(r_j^2 + r_{j+1}^2\right)}{E_{j+1}\left(r_{j+1}^2 - r_j^2\right)} + \dfrac{b_{j+1}\,\nu_{j+1}}{E_{j+1}}$;

or $B_{j;\,j+1} = -\dfrac{A_{in;\,j} + A_{out;\,j} + A_{in;\,j+1} + A_{out;\,j+1}}{2} - \dfrac{b_j\,\nu_j}{E_j} + \dfrac{b_{j+1}\,\nu_{j+1}}{E_{j+1}}$; for axisymmetric plane stress state $a_j = b_j = 1$ and for axisymmetric plane strain state $a_j = 1 - \nu_j^2$ and $b_j = 1 + \nu_j$.

Combination all (6) equations written for all cylinder layers, gives a system of linear equations which solution are contact pressures p_1, p_2, …, p_{n-1}:

$$\begin{bmatrix} B_{1;2} & A_{out;2} & 0 & \cdots & 0 & 0 & 0 \\ A_{in;2} & B_{2;3} & A_{out;3} & \cdots & 0 & 0 & 0 \\ 0 & A_{in;3} & B_{3;4} & \cdots & 0 & 0 & 0 \\ \vdots & \vdots & \vdots & \ddots & \vdots & \vdots & \vdots \\ 0 & 0 & 0 & \cdots & B_{n-3;n-2} & A_{out;n-2} & 0 \\ 0 & 0 & 0 & \cdots & A_{in;n-2} & B_{n-2;n-1} & A_{out;n-1} \\ 0 & 0 & 0 & \cdots & 0 & A_{in;n-1} & B_{n-1;n} \end{bmatrix} \begin{Bmatrix} p_1 \\ p_2 \\ p_3 \\ \vdots \\ p_{n-3} \\ p_{n-2} \\ p_{n-1} \end{Bmatrix} = \begin{Bmatrix} \Delta_{1;2}/r_1 - A_{in;1}\, p_0 \\ \Delta_{2;3}/r_2 \\ \Delta_{3;4}/r_3 \\ \vdots \\ \Delta_{n-3;n-2}/r_{n-3} \\ \Delta_{n-2;n-1}/r_{n-2} \\ \Delta_{n-1;n}/r_{n-1} - A_{out;n}\, p_n \end{Bmatrix}. \tag{7}$$

Strain state components in n-layer pipe can be obtain by using generalized Hooke's law.
Stress and strain intensity (*von Mises* criteria) at axisymmetric plane stress state:

$$\sigma_i = \sqrt{\sigma_r^2 + \sigma_\theta^2 - \sigma_r\,\sigma_\theta}\,; \tag{8}$$

$$e_i = \frac{\sigma_i}{E} = \frac{1}{\sqrt{2}\,(1+\nu)}\sqrt{(e_r - e_\theta)^2 + (e_\theta - e_z)^2 + (e_z - e_r)^2}\,; \tag{9}$$

and at axisymmetric plane strain state:

$$\sigma_i = \sqrt{(\sigma_\theta - \sigma_r)^2 - \nu(1-\nu)(\sigma_\theta + \sigma_r)^2 + \sigma_r\,\sigma_\theta}\,; \tag{10}$$

$$e_i = \frac{\sigma_i}{E} = \frac{1}{1+\nu}\sqrt{e_r^2 + e_\theta^2 - e_r\,e_\theta}\,. \tag{11}$$

In 2nd chapter presented method for determination of stress strain state components in multilayer cylinder subjected to internal and/or external pressure is valid if materials of cylinder layers are deformed only elastically. By using von Mises yield criteria above presented method is valid then

$$\sigma_i \le \sigma_y. \tag{12}$$

2. Validation of presented method

The accuracy of the presented method was verified by the finite element method. The *Ansys Student 2020 R2* program was used for this. The multilayer cylinder model was meshed by PLANE183 type finite elements. The size of the finite elements was chosen so that there are at least 50 elements along the wall thickness of the multilayer cylinder, but not less than 4 elements along the thickness of any layer. TARGE169 (assigned to the contacting surface of the layer with the stiffer material) and CONTA172 finite elements were used for the contact modeling. The contact problem was solved using the augmented Lagrange method and default program settings.

As an example, the stress strain state in 5-layer hollow cylinder was analyzed. The data of 5-layer hollow cylinder are given in Table 1. For this cylinder (7) will acquire the following form

$$\begin{bmatrix} B_{1;2} & A_{out;2} & 0 & 0 \\ A_{in;2} & B_{2;3} & A_{out;3} & 0 \\ 0 & A_{in;3} & B_{3;4} & A_{out;4} \\ 0 & 0 & A_{in;4} & B_{4;5} \end{bmatrix} \begin{Bmatrix} p_1 \\ p_2 \\ p_3 \\ p_4 \end{Bmatrix} = \begin{Bmatrix} \Delta_{1;2}/r_1 - A_{in;1}\, p_0 \\ \Delta_{2;3}/r_2 \\ \Delta_{3;4}/r_3 \\ \Delta_{4;5}/r_4 - A_{out;5}\, p_5 \end{Bmatrix}.$$

Table 1. The thickness and material mechanical properties of layers of 5-layer hollow cylinder Ø16 × 2

Cylinder layer number	Layer thickness, mm	Elastic modulus E of layer material, MPa	Poisson's ratio v of layer material
1	0.40	650	0.42
2	0.15	800	0.40
3	0.6	1000	0.38
4	0.35	800	0.40
5	0.5	650	0.42

Comparison of stress strain components obtained analytically and determined by FEA are presented in Figures 3–6. The largest disagreement for stress components does not exceed 0.5 % and for strains components – 1.0 %.

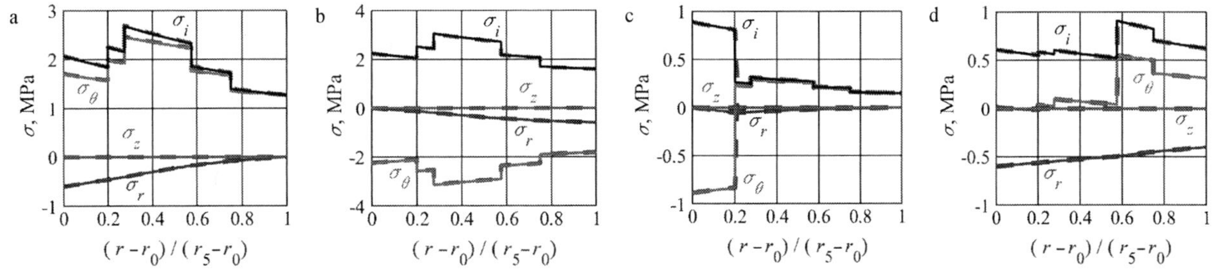

Fig. 3. Distribution of stress state components in the thickness of 5-layer cylinder wall under axisymmetric plane stress state determined analytically /——/ and by FEA / – – – /: (a) subjected only to inner pressure $p_0 = 0.6$ MPa; (b) subjected only to external pressure $p_5 = 0.6$ MPa; (c) without inner and external pressures, with interference of 1st and 2nd layers by 10 μm; (d) subjected to inner $p_0 = 0.6$ MPa and external $p_5 = 0.4$ MPa pressures, with interference of 3rd and 4th layers by 5 μm

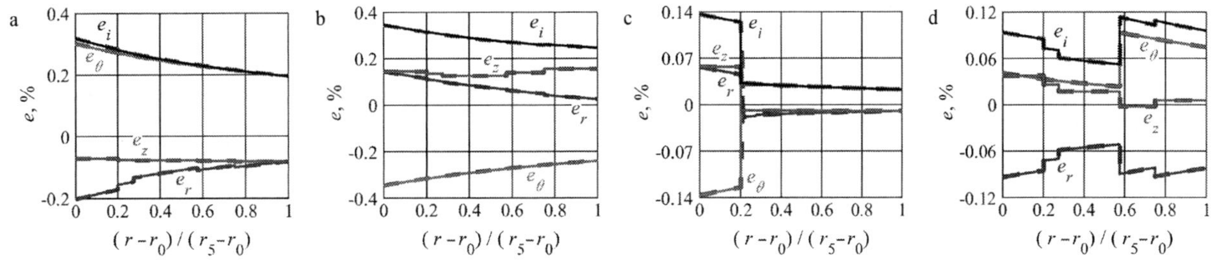

Fig. 4. Distribution of strain state components in the thickness of 5-layer cylinder wall under axisymmetric plane stress state. Other data is the same as in Figure 3

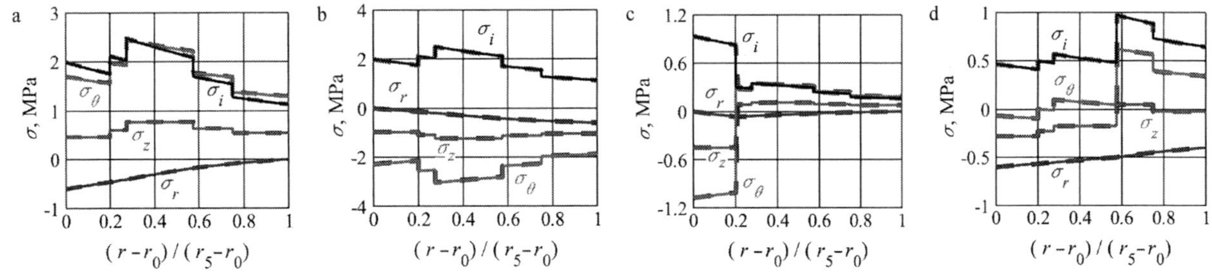

Fig. 5. Distribution of stress state components in the thickness of 5-layer cylinder wall under axisymmetric plane strain state. Other data is the same as in Figure 3

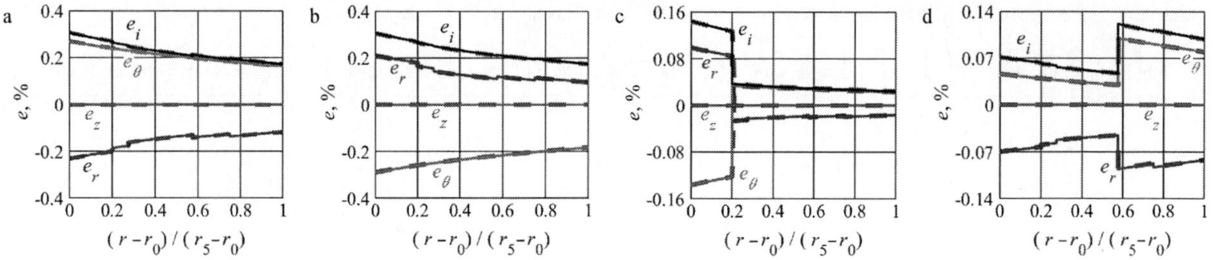

Fig. 6. Distribution of strain state components in the thickness of 5-layer cylinder wall under axisymmetric plane strain state. Other data is the same as in Figure 3

Conclusion

The analytical method for determination of stress and strain components in multilayer cylinder subjected to hydrostatic inner or / and external pressure under elastic loading and axisymmetric plane stress or strain state is presented in this paper. By FEA it is proved that accuracy of this method is good.

References

[1] Patil, J. S., Ingale, S. M. Study of multilayered composite pipe subjected with metal interlayer. *International Research Journal of Engineering and Technology (IRJET)*; 5(1):1465–1467.

[2] Sülü, I. Y., Temiz, S. Stress analyses of multi-layered composite pipes subjected to internal pressure. *Academic Journal of Science*, 2015; 4(03):187–194.

[3] Shildip, D. U., Bhope, D. V., Khamankar, S. D. Stress analysis of multilayer pressure vessel. *Journal of Applied oJ Mechanical Engineering*. 2015; 4(2):157. doi:10.4172/2168-9873.1000157.

[4] Trifonov, O. N., Cherniy, V. P. Analysis of stress-strain state in steel pipe strengthened with a composite wrap. *Journal of Pressure Vessel Technology*, 2014; 136(5):051202.

[5] Mandalapu, T., Krishnamoorthy, S. R. Design and analysis of multilayer high pressure vessels and piping. *IJSRD – International Journal for Scientific Research & Development*, 2015; 3(2):1076–1083. ISSN (online): 2321-0613.

[6] Feodosjev, V. *Medžiagų atsparumas*. Vilnius: Mokslas, 1977; 524 p.

[7] Eraslan, A. N., Akiş, T. Deformation analysis of elastic-plastic two layer tubes subject to pressure: an analytical approach. *Turkish Journal of Engineering and Environmental Sciences*, 2004, pp. 261–268.

[8] Qiu, J., Zhou, M. Analytical solution for interference fit for multi-layer thick-walled cylinders and the application in crankshaft bearing design. *Applied Sciences*, 2016; 6(6):167; doi:10.3390/app6060167.

[9] Bai, Z., Wang, J., Ning, K., Hou, D. Contact pressure algorithm of multi-layer interference fit considering centrifugal force and temperature gradient. *Applied Sciences*, 2018; 8(5):726. doi:10.3390/app8050726.

Author Index